IET ENERGY ENGINEERING 184

Compressed Air Energy Storage

Other volumes in this series:

Compressed Air Energy Storage

Types, systems and applications

Edited by
David S-K. Ting and Jacqueline A. Stagner

The Institution of Engineering and Technology

Published by The Institution of Engineering and Technology, London, United Kingdom

The Institution of Engineering and Technology is registered as a Charity in England & Wales (no. 211014) and Scotland (no. SC038698).

© The Institution of Engineering and Technology 2021

First published 2021

The Institution of Engineering and Technology
Michael Faraday House
Six Hills Way, Stevenage
Herts, SG1 2AY, United Kingdom

www.theiet.org

British Library Cataloguing in Publication Data
A catalogue record for this product is available from the British Library

ISBN 978-1-83953-195-8 (Hardback)
ISBN 978-1-83953-196-5 (PDF)

Typeset in India by MPS Limited

Contents

About the editors

David S-K. Ting is a professor in Mechanical, Automotive and Materials Engineering and the founder of the Turbulence & Energy Laboratory at the University of Windsor, Canada. His memberships include ASHRAE, ASME, and SAE. He has authored or co-authored more than 150 journal papers, 4 textbooks, and co-edited 14 books.

Jacqueline A. Stagner is a professional engineer, and the Undergraduate Programs Coordinator in the Faculty of Engineering at the University of Windsor, Canada. As an active member of the Turbulence & Energy Laboratory, she supervises students in sustainable energy and technologies, and has co-edited related books.

Preface

In Chapter 1, Ting and Stagner introduce the purpose of this volume, disseminating the latest advancements in compressed air energy storage (CAES) and its future perspectives. Next, Mehri acquaints the reader with CAES in Chapter 2. It is noted that CAES is not a mature technology; that is, there is a lot of room for further improvement. This is followed by a chapter dedicated to isothermal CAES, Chapter 3, put together by Zhang, Chen, Xu, Zhou and Guo, being able to effectively capture and release thermal energy is critical in approaching isothermal status. In Chapter 4, Ebrahimi, Brown, Ting, Carriveau and McGillis discuss preconditioning of the discharge air stream to improve the performance of adiabatic CAES. In particular, they analyse the effects of air temperature and humidity. Li and Lin bring us up to date on the large-scale commercialization of CAES in Chapter 5. They focus on the technical issues associated with underground reservoirs. Included in this chapter is a case study where a salt cavern in Ontario, Canada, is employed for CAES. Beyond providing an overview of CAES in Chapter 6, Salvador, Schmitz, Lazzarin and Coelho focus on small-scale CAES. They talk about hybridization to further CAES and renewable energy. Hybridization is equally indispensable for successful large-scale CAES, according to Llamas, Blanco-Brox, Castaneda and Barthelemy, as detailed in Chapter 7; CAES can be complemented with pumped hydro, hydrogen energy storage or biomass gasification. In Chapter 8, Ke, He, Dooner, Luo and Wang update the reader concerning dynamic modelling of CAES. This chapter is based on extensive research performed at the University of Warwick. Proper scheduling is essential to better and further embrace renewable energy enabled by CAES. This is detailed in Chapter 9 by Hlalele, Zhang, Naidoo and Bansal; the key is to participate in the day-ahead market. We would not do justice to this volume without talking about the driver, money. For money to talk in greening energy, Steeb and Xydis recommend potent policies to promote renewable energy investments in Chapter 10. Next, by an exergy analysis of a small-scale trigenerative CAES (T-CAES) in Chapter 11, Dittakavi, Ting, Carriveau and Ebrahimi demonstrate the utility of exergy analysis in identifying specific areas for furthering T-CAES. The volume concludes with 'Offshore systems' by Onwuchekwa as Chapter 12. The vast waters furnish much room for harnessing renewable energy and thus its better half CAES.

Acknowledgements

The editors are most grateful to Dr Christoph von Friedeburg and IET, especially Olivia Wilkins and Paul Deards, for trusting and helping us to realize this timely volume. We thank the experts who compiled the 12 chapters. It is a privilege to join them in savouring the fruits of their labour. A big shout-out goes to the distinguished but anonymous reviewers who heightened the quality of each manuscript. Lastly, grace from above sustained this project through the entire journey.

Chapter 1

The current status and future perspectives of compressed air energy storage

David S-K. Ting[1] and Jacqueline A. Stagner[1]

It is trite to say that energy storage is essential for furthering renewable energy by stabilizing the supply and demand. It is also cliché to point out that compressed air energy storage (CAES) is a promising means for energy storage. To highlight but a few of the multitude of recent publications on CAES, Tan *et al.* [1] present a comprehensive review concerning various energy storage technologies for empowering smart grid. CAES is also one of the most promising energy storage means in the harsh marine environment [2]. Guo *et al.* [3] discuss the promise and challenges of utility-scale CAES in aquifers. A regional review of CAES for northern China is compiled by Tong *et al.* [4]. Mahmoud *et al.* [5] compare and contrast the three main mechanical energy storage options, flywheel, pumped hydro, and CAES. They conclude that flywheel is best suited for short-duration applications. For longer durations, pumped hydro has the efficiency while CAES provides a faster start-up. For good environmental stewardship, adiabatic or isothermal CAES is recommended. In short, the case for CAES is clear. It is expected to ramp up its importance as we march forward to harness progressively greener energy.

Keywords: Compressed air energy storage; renewable energy; smart grid

1.1   An introduction to CAES

The next chapter conveys a review of the general notion of energy storage. Namely, Mehri briefs the reader with a succinct historic timeline of CAES, in Chapter 2. He then talks about the main components associated with CAES. Included are the three types of air storage: isochoric, isobaric, and cryogenic. The main types of expanders appropriate for recovering energy from compressed air are also discussed. Expounded in detail are the three classes of CAES, diabatic, adiabatic, and

---

[1]Turbulence and Energy Laboratory, University of Windsor, Windsor, Ontario, Canada

Table 1.1   *The three classes of CAES*

CAES systems			
Diabatic		**Adiabatic**	**Isothermal**
D-CAES	A-CAES with TES	A-CAES without TES	I-CAES

TES, thermal energy storage.

isothermal, as summarized in Table 1.1. Also covered are newer CAES technologies such as supercritical CAES, underwater CAES, and offshore CAES. After a concise discussion of CAES risk assessment, the chapter closes with an outlook for CAES, noting that CAES is well suited for managing the fluctuation of wind power.

1.2   Isothermal CAES

Chapter 3 is a thorough chapter on isothermal CAES (I-CAES) written by a group of experts who have been intensely involved with frontier experimental and analytical research on adiabatic and, now, isothermal CAES. Zhang, Chen, Xu, Zhou, and Guo explain that being able to effectively capture and release thermal energy is critical to approaching the isothermal status. After elucidating the working principle of I-CAES, the research progress of I-CAES is conveyed. They propose a practically achievable quasi-isothermal expansion process by injecting water into the expander cylinder. Their model is validated with empirical results from a novel three-stage reciprocating adiabatic expander they developed recently. The simulations show that spraying tiny water droplets into the cylinder can improve the specific work output beyond that made possible by an adiabatic expander. For the study's conditions, operating the expander under quasi-isothermal mode leads to a cylinder height reduction of 8.7%, specific work output increase of 15.7%, and inlet/outlet temperature difference of only 10% of that in the adiabatic expansion process. These are extremely encouraging results, lending promise to a significant boost in commercial CAES performance in the near future.

1.3   Adiabatic CAES

Adiabatic CAES (A-CAES) is another class of CAES that deserves better understanding, and its potential is far from being fully exploited in practice, as discussed in Chapter 4. The idea is to recover heat generated during the compression phase and to put it back into the system during the expansion phase. In this way, a significant amount of energy can be retained within the charging–discharging process, leading to a substantial increase in efficiency. Ebrahimi, Brown, Ting, Carriveau, and McGillis argue that preconditioning of the discharge air stream can result in a noticeable improvement in the performance of A-CAES. They thermodynamically simulated a 100 MW A-CAES system where the cavern pressure is varied between 4 and 8 MPa. The maximum attainable efficiency is found to be over 80%.

1.4 Technical feasibility analysis of underground reservoir CAES

Li and Lin bring us up to date with large-scale commercialization of CAES, in Chapter 5. They focus on the technical issues associated with the underground reservoir. Among a handful of operating CAES facilities around the world, there are close to a handful currently under construction, and more in the planning stage. This reality alone is more than enough to convince us of what the future holds for CAES. Due to proximity, Li and Lin performed their case study on a salt cavern in Ontario, Canada. They draw their chapter to a close reiterating the fact that CAES furnishes an effective means to handle power fluctuations and uncertainty of renewable energy. They highlight the need to assess the geological and technical issues of the potential underground reservoir.

1.5 Small-scale CAES

Chapter 6 starts with an overview on CAES; Salvador, Schmitz, Lazzarin, and Coelho underscore the necessity of sound thermal management through adiabatic and isothermal processes. The focus of the chapter is small-scale CAES (SS-CAES). SS-CAES systems have been employed as an alternative to replace batteries in autonomous and distributed generation systems. SS-CAES with pneumatic microturbines generally operate at lower pressure levels compared to the large-scale CAES. As a result, the efficiency is reduced. Salvador *et al.* point out that hybrid SS-CAES systems, such as hybrid compressed air and supercapacitors and battery with oil-hydraulics pneumatics (BOP), have definite advantages under certain conditions. The point is to tap into the better efficiency of hydraulic motors and pumps, along with their higher pressure levels. The combination, however, requires an interface between fluids, oil, and air, making the system more complex. They evaluate a couple of different BOP systems thoroughly. These applications call for reliable control strategies that can track down the optimal operation conditions, as the load and other parameters such as pressure change.

1.6 Hybridization of large-scale CAES

Hybridization is equally indispensable for successful large-scale subsurface CAES, according to Llamas, Blanco-Brox, Castaneda, and Barthelemy, as disclosed in Chapter 7. CAES can be complemented with pumped hydro, hydrogen energy storage, or biomass gasification. To better capitalize this in practice, we should take advantage of obsolete infrastructure, along with a novel alternative to thermal energy management. Llamas *et al.* tabulate a business case comparison between pumped hydro, CAES, and hydrogen fuel cells, highlighting the competitiveness of CAES. The case for hybridization is illustrated using hybrid CAES and hydrogen energy storage, hybrid CAES and low-temperature thermal energy storage, hybrid CAES and high-temperature system, hybrid CAES and biomass gasification or biogas, and others. They advocate that CAES should coexist with other mass energy storage systems.

1.7 Dynamic modeling of CAES

In Chapter 8, Ke, He, Dooner, Luo, and Wang update the reader concerning dynamic modeling of CAES. The advancement of model simulations has been resolving many challenges associated with CAES. Various models can be used for design, optimization, control, and implementation of CAES. This chapter is based on extensive research performed at the University of Warwick. Ke *et al.* describe and evaluate five types of simulation models for CAES systems and components, namely, thermodynamic models, empirical models, data-driven models, lumped-parameter models, and distributed-parameter or full geometry-based models. The system description ability and computational cost for these types of models are compared. The selection depends on the specific purpose. After delineating the mathematical details underlying the expander, heat exchanger, thermal energy storage, and generator, a novel data-driven dynamic simulation approach for a complex system is demonstrated in detail. They show that this innovative neural network model can significantly reduce computational cost and potentially be able to help build CAES system-level, real-time simulations.

1.8 CAES for day-ahead dispatch scheduling of renewable energy

Proper scheduling is the key to better and further embrace renewable energy enabled by CAES. The doorway is to participate in the day-ahead market. This is explicated in Chapter 9 by Hlalele, Zhang, Naidoo, and Bansal. They present a combined day-ahead dispatch schedule for CAES systems with renewable energy sources under demand response and renewable obligation. CAES is exploited to overcome the uncertainty associated with wind and solar energy. Hlalele *et al.* apply the model to a renewable obligation policy, guaranteeing a defined portion of the energy is from renewable energy sources. CAES systems are integrated with renewable energy source generators and demand response is used for deferring flexible demand from peak-electricity-price periods to low-price periods. Real large-scale data that allow the system operator to manipulate the participation of electric water heaters from the substation level are analyzed. The model is tested on a modified IEEE 30-bus system that has 6 thermal units and 41 transmission lines. A benchmark of 10% renewable energy penetration level is used. The results exhibit that CAES can increase renewable energy penetration and profitability.

1.9   Direct air capture and wind curtailment

We would not do justice to this volume without a chapter dedicated to the driver, money. For money to talk in greening energy, Steeb and Xydis recommend potent policies to promote renewable energy investments, in Chapter 10. The idea is to reduce waste due to wind curtailment by integrating higher amounts of renewable energy sources, in addition to modernizing the grids. Also needed are incentives called negative prices, a business model that is employee-unfriendly in the sense that the companies have to frequently fire their workers in order to remain profitable. A viable solution is direct air capture, that is, making use of the otherwise curtailed or unwanted wind energy on-site for direct air carbon capture and storage.

1.10   Exergy analysis of a small-scale trigenerative CAES

An exergy analysis of a small-scale trigenerative CAES (T-CAES) is furnished as Chapter 11. The key to trigeneration is the direct usage of the relatively low-quality heat. Dittakavi, Ting, Carriveau, and Ebrahimi demonstrate the utility of exergy analysis in identifying specific areas for furthering T-CAES. The studied T-CAES system has a 4 kW compressor and a 2 kW turbine. It is found that more than half of the total exergy destruction is associated with the accumulator. The next largest destroyers are the pressure regulator, followed by the turbine. The take-home message is that there is great potential for further improvement in the studied T-CAES and, also, CAES, in general.

1.11   Offshore CAES

The volume concludes with "Offshore systems" by Onwuchekwa, in Chapter 12. The vast waters provide much room for harnessing renewable energy and, thus, also its companion, CAES. The expander in this particular case is a rack and pinion gear system that taps into the torque powered by the rising energy bag. This energy bag or balloon is better known as underwater CAES (UW-CAES). The torque created by the buoyancy force is transferred through a shaft to an electric generator. The advantages of this system, instead of expanding the compressed air through a conventional expander, include quick start-ups and variable speeds. For the considered conditions, the design calculations indicate that a high torque at a low angular speed can lead to definite power output. Onwuchekwa also suggests that a purpose-built electric arc furnace can be designed to heat molten salts to over 550 °C. This heat can be utilized, along with the harnessed heat of compression, to enhance the work output.

1.12   CAES outlook

Though far from exhaustive, this volume echoes and authenticates that, in recent years, CAES has rightly claimed an important status in enabling renewable energy, it positions itself to ramp up the greening of tomorrow. It is clear that there is much room for ingenuity within the umbrella of CAES. Moreover, there are also many promising opportunities for integrating CAES with other innovative systems. The advancements in CAES are in everybody's best interest. We look forward to providing another forum to stimulate discussion and dissemination of tomorrow's advances in CAES.

References

[1] K. M. Tan, T. S. Babu, V. K. Ramachandaramurthy, P. Kasinathan, S. G. Solanki, and S. K. Raveendran, "Empowering smart grid: A comprehensive review of energy storage technology and application with renewable energy integration," *Journal of Energy Storage*, 39: 102591, 2021.

[2] Z. Wang, R. Carriveau, D.S-K. Ting, W. Xiong, and Z. Wang, "A review of marine renewable energy storage," *International Journal of Energy Research*, 43(12): 6108–6150, 2019.

[3] C. Guo, C. Li, K. Zhang, *et al.* "The promise and challenges of utility-scale compressed air energy storage in aquifers," *Applied Energy*, 286: 116513, 2021.

[4] Z. Tong, Z. Cheng, and S. Tong, "A review on the development of compressed air energy storage in China: Technical and economic challenges to commercialization," *Renewable and Sustainable Energy Reviews*, 135: 110178, 2021.

[5] M. Mahmoud, M. Ramadan, A-G. Olabi, K. Pullen, and S. Naher, "A review of mechanical energy storage systems combined with wind and solar applications," *Energy Conversion and Management*, 210: 112670, 2020.

Chapter 2

An overview of CAES

Mohsen Mehri[1]

Energy storage systems as a part of energy secure supply have the ability to take up a certain amount of energy, store it in a storage medium for a suitable period of time, and release it in a controlled manner after a certain time delay. Large-scale mechanical storage of electric power is currently almost exclusively achieved by pumped-storage hydroelectric power stations. In the area of electrochemical storage, different technologies are currently in various stages of research, development, and demonstration of their suitability for large-scale electrical energy storage. Thermal energy storage technologies are based on the storage of sensible heat, exploitation of phase transitions, adsorption/desorption processes, and chemical reactions. In thermo-mechanical energy storage systems like compressed air energy storage (CAES), energy is stored as compressed air in a reservoir during off-peak periods, while it is used on demand during peak periods to generate power with a turbo-generator system. In the following chapter, after introduction of system key components, timeline development and progress of CAES from different point of view is discussed. Such plants can offer significant benefits in terms of flexibility in matching a fluctuating power demand, particularly when coupled with renewable sources. Storing energy when it's made and releasing it when it's needed helps keep the grid reliable and paves the way for introducing intermittent renewables like wind and solar to the mix. CAES is NOT "a mature technology." Also, it is not "a single technology." There is huge scope for further learning and further cost reduction.

Keywords: Energy storage; compressed air; thermal energy storage; thermodynamic categorization

[1]Faculty of Mechanical Engineering, College of Engineering, University of Tehran, Tehran, Iran

Nomenclature

A-CAES	Adiabatic compressed air energy storage
AA-CAES	Advanced adiabatic compressed air energy storage
AEC	Alabama Electric Corporation
C-CAES	Conventional compressed air energy storage
CAES	Compressed air energy storage
CAS	Compressed air storage
CHP	Combined heat and power
D-CAES	Diabatic compressed air energy storage
DOE	Department of Energy
DS	Specific diameter
EHS	External heat source
HP	High pressure
I-CAES	Isothermal compressed air energy storage
IEHS	Integrated electricity and heating systems
LAES	Liquid air energy storage
LHS	Latent heat storage
MRE	Marine renewable energy
NS	Specific speed
OCAES	Offshore compressed air energy storage
PCM	Phase change materials
PNNL	Pacific Northwest National Laboratory
RES	Renewable energy source
SHS	Sensible heat storage
SC-CAES	Supercritical compressed air energy storage
TES	Thermal energy storage
UWCAES	Underwater compressed air energy storage

2.1   Introduction: motivation and principles

In manufacturing industry, compressed air is broadly applied—it is used either as an energy carrier or it serves as a process fluid carrier. In Germany, for example, currently 16 TWh of electricity is consumed annually to provide compressed air for industrial purposes, which amounts to about 2.5% of the German overall electricity consumption. The fundamental idea to store electrical energy by means of compressed air dates back to the early 1940s. By then, the patent application "Means for storing fluids for power generation" was submitted by F.W. Gay to the US Patent Office. In 1969, the need for storage capacity in northern Germany led to the decision to develop a CAES plant in this particular region. Stimulated by the

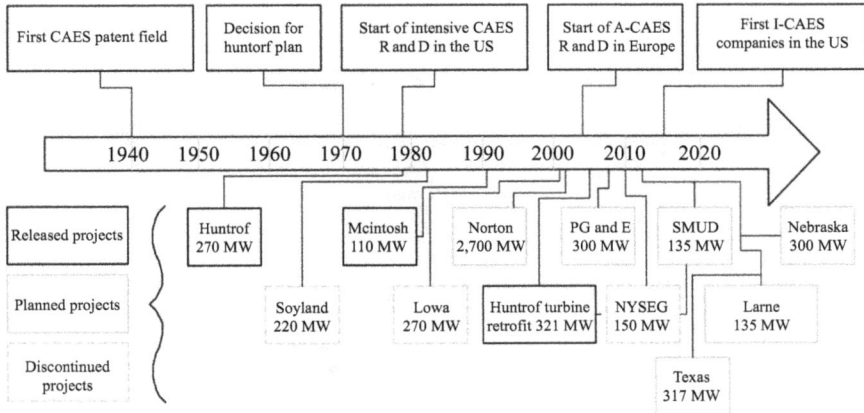

Figure 2.1   Timeline of CAES R&D and industrial efforts

Huntorf project, the general interest in CAES technology began to rise by the mid-1970s by DOE and initiated both an R&D and a pre-demonstration program for developing CAES, which was coordinated by the Pacific Northwest National Laboratory (PNNL) from the late 1970s to early 1980s. At the end of the research program, diabatic CAES (D-CAES) was considered a technically feasible near-term technology. Development of second generation CAES like hybrid, adiabatic, or isothermal CAES (I-CAES) was postponed and linked to a successful implementation of D-CAES in the USA. The first CAES plant in the USA was actually built in 1991 at the McIntosh site. Figure 2.1 shows projects and R&D efforts over time, and the very beginning of the twenty-first century can be seen as the point in time when R&D on CAES technology has been resumed on a broader level. All different types of CAES plant concepts known today have their origin in this decade. Replacement of current gas turbines in the USA by CAES plants could result in annual savings of more than 100,000,000 barrels of oil. The net annual oil saving would be greater in the future. CAES plants are not limited to the same degree by the siting difficulties faced by conventional pumped hydro plants. Finally, a well-designed CAES plant should have a smaller adverse impact on the environment compared to a conventional gas turbine peaking plant.

2.2   Key components

The basic concept of CAES is rather simple. The storage is charged by the use of electrically driven compressors which convert the electric energy into potential energy, or more precisely exergy, of pressurized air. The pressurized air is stored in compressed air storage (CAS) volumes of any kind and can then be released upon demand to generate electricity again by expansion of the air through an air turbine. Actually, the operation principle of CAES facility is almost similar to the

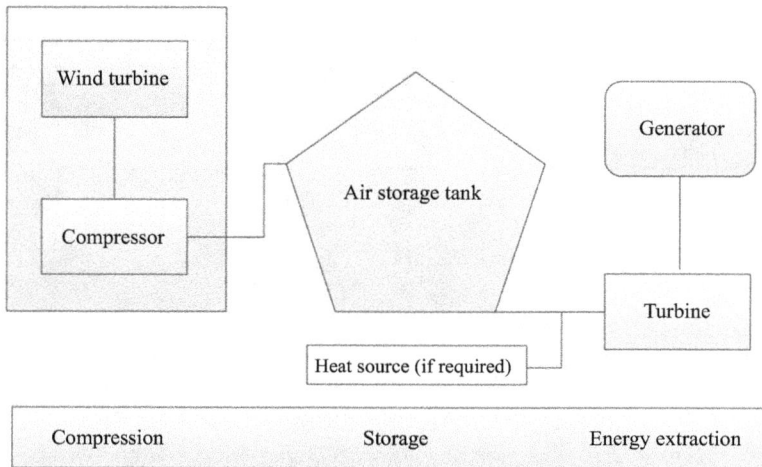

Figure 2.2   Process of CEAS-based power generation

conventional gas turbine. The main difference is the separation of compression process and expansion process in CAES system. Compressor, air storage reservoir, and expander are the three main components in CAES system. CAES is one of the promising methods for the combination of renewable energy source (RES)-based plants with electricity supply, and has a large potential to compensate for the fluctuating nature of renewable energies. Schematic for such combination is shown in Figure 2.2.

Compression machinery is very well established in the range of 1–12 bars. Modern aero engines have axial compressors with pressure ratio more than 50. The compressor used in CAES is to inject air from ambient into the air storage reservoir in a high-pressure (HP) status by using the off-peak electricity. The main sources of energy loss in compressor are the compression heat and the mechanical conversion loss. Low-pressure machines are the most expensive per kilowatt. Different compressor technology versus flow rate and pressure delivered are shown in the graph of Figure 2.3

The air storage reservoir is in charge of storing air in charge process and providing air in discharge process. The energy loss in this component is mainly caused by the heat dissipation through the reservoir wall. Compressed air can be stored either at constant volume (isochoric) or at constant pressure (isobaric).

In case of constant volume storage, the pressure varies and thus indicates the state of charge. The most common example of isochoric storage is a steel pressure vessel or, at large scale, a salt cavern. Salt caverns make good energy storage reservoirs as they are impermeable and do not react with oxygen.

Constant pressure storage requires a varying volume to maintain pressure at a constant level while charging and discharging. In this case, the volume indicates the state of charge. Constant pressure storage can technically be realized using a

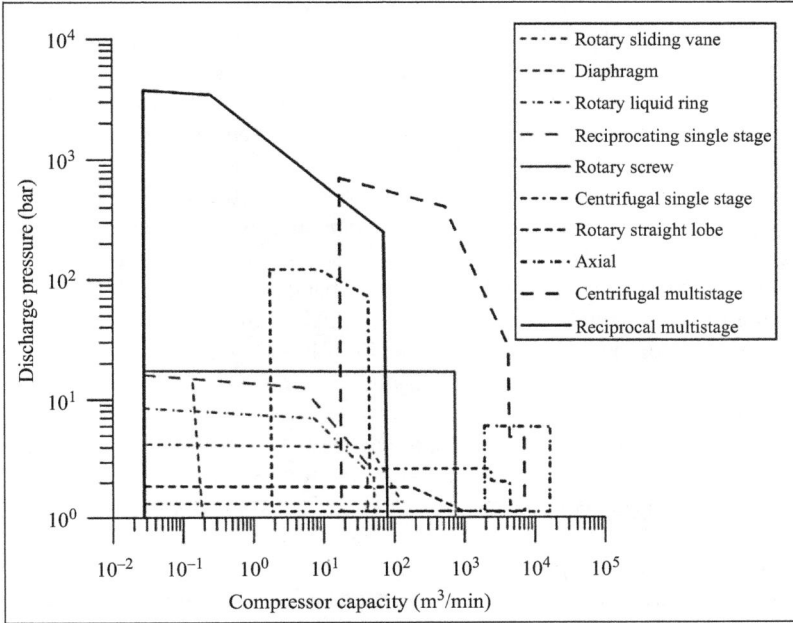

Figure 2.3   Typical operating range of different compressor type [1]

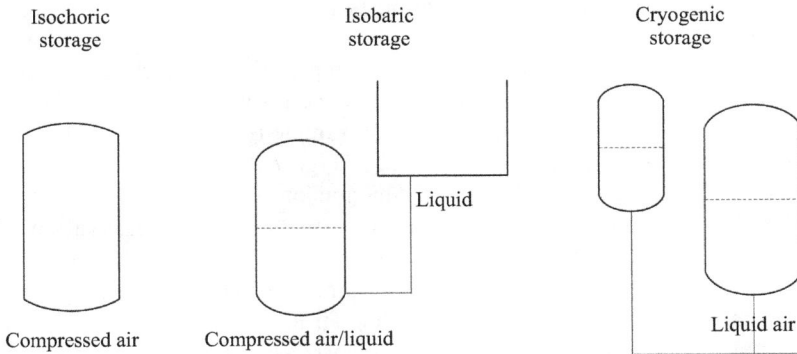

Figure 2.4   Different types of air storage devices

hydraulically compensated reservoir where pressure is kept approximately constant by a second reservoir of liquid at elevated geodetic height as depicted in the center of Figure 2.4.

Beside the isochoric and isobaric storage of compressed air, there is also the possibility to store the air as a liquid at cryogenic temperatures. The use of cryogenic storage requires a change in energy conversion technology as well. This so-called liquid air energy storage (LAES) technology is not only related to CAES but also to air separation facilities. LAES layouts can be subdivided in diabatic, adiabatic, and isothermal processes, just like CAES layouts. As the focus of this chapter is on CAES technology, LAES is mentioned just for the sake of completeness.

The power conversion machinery and the HP air stores are characterized by very long natural lifetimes.

The expander in CAES system is used to convert the stored energy to electricity during peak period. Generally, the expander can be divided into two main categories, the volumetric expanders and the turbomachines. The volumetric expanders contain the scroll expander, screw expander, and reciprocating expander. The turbomachines include axial turbine and radial turbine. The main energy losses occurred in this component is heat dissipation of high-temperature flue gas and the mechanical conversion efficiency of expander. Information data about the turbines are shown in Table 2.1.

For a given volume flow rate and a given head change through a turbomachine, specific speed (NS) is a number indicative of the rotative speed of the machine and specific diameter (DS) is a number indicative of the rotor diameter or size of the machine.

Reynolds number expresses the ratio of inertia force to viscous force and reflects the properties of the fluid and the speed of the machine.

Suction NS for machines such as pumps operating on noncompressible fluids will indicate whether or not cavitation exists. If cavitation does not exist, then pump performance will be as expected. If serious cavitation exists, the pump performance cannot be predicted from the similarity parameters. For machines operating on compressible fluids such as turbines and compressors, Mach number is used as the fourth similarity parameter in place of suction NS. If the Mach number of the machine is less than or near 1.0, the compressibility effects are negligible, which eliminates this variable and turbomachine performance can be presented as a function of two parameters, NS and DS. An example of a typical NS–DS correlation for full admission turbines is presented in Figure 2.5.

The performance of turbines as a function of NS and DS is essentially a solution of the theoretical hydraulic efficiency equation and the calculation of the parasitic losses.

CAES includes a broad set of exergy storage technologies, which involves compressing air to store exergy and expanding air to release exergy. CAES always involves managing heat as well as HP air. In most cases of CAES systems, the cost of the HP air store dominates the cost of the system. Today, a huge variety of different CAES concepts exist at different levels of development, aiming at different applications and owning individual strengths and weaknesses. A general classification of the whole group of CAES concepts is shown in Figure 2.6

Depending on the targeted idealized process, CAES technologies are differentiated into diabatic, adiabatic, and isothermal concepts. The main criterion for categorization is the question how heat is handled during compression and prior to expansion of the air. In D-CAES, the heat resulting from air compression is wasted to the ambient by cooling down the compressed air; therefore, an external heat source

Table 2.1 Information data about the turbines [2]

Type	Rotate speed (rpm)	Cost	Advantages	Disadvantages
Radial-inflow turbine	8,000–80,000	High	Light weight Mature manufacturability and high efficiency	High cost Low efficiency in off-design conditions and cannot bear two-phase
Scroll expander	<6,000	Low	High efficiency Simple manufacture Light weight Low rotate speed and tolerable two-phase	Low capacity Lubrication and modification requirement
Screw expander	<6,000	Medium	Tolerable two-phase Low rotate sped and high efficiency in off-design conditions	Lubrication requirement Difficult manufacture and seal
Reciprocating piston expander	–	Medium	HP radio Mature manufacturability Adaptable in variable working condition and tolerable two-phase	Many movement parts Difficult manufacture and seal
Rotary vane expander	<6,000	Low	Tolerable two-phase Torque stable Simple structure Low cost and noise	Lubrication requirement and low capacity

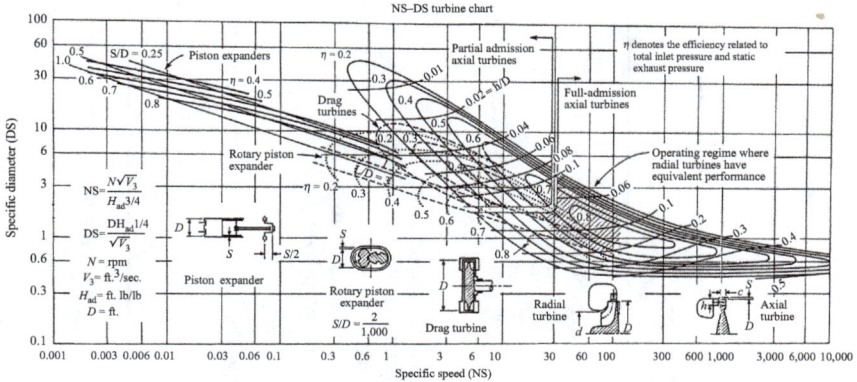

Figure 2.5   Typical NS–DS correlation for full-admission turbines [3]

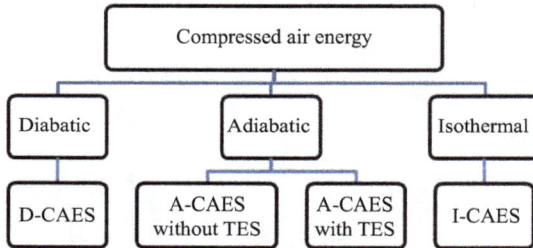

Figure 2.6   CAES concepts classified by their idealized change of state:
D (diabatic), A (adiabatic), and I (isothermal) CAES

(EHS) is needed for the discharging process to prevent condensation in and icing of the expansion machinery by preheating the compressed air upstream of the expander.

In adiabatic CAES (A-CAES), the heat of compression is captured in additional thermal energy storage (TES) devices and is utilized prior to expansion to prevent the need for other heat sources during the discharge phase. In contrast to D-CAES and A-CAES concepts, heat of compression is to be minimized or even prevented in I-CAES concepts.

2.3   D-CAES systems

Existing grid-scale plants are diabatic; they use fuel to heat the air prior to expansion of the stored air. D-CAES is not really CAES at all; ~50% of the exergy in existing grid-scale CAES systems is stored in the fuel. So far, there are only two CAES plants in the world:

The 290 MW plant belonging to E.ON Kraftwerke, Huntorf, Germany, built in 1978, which is designed to provide black start service to a nuclear plant initially and the 110 MW plant of Alabama Electric Corporation (AEC) in Mcintosh, Alabama, USA, commissioned in 1991 which can provide the electricity for 26 h at full power condition. Both plants are of the D-CAES type, use solution-mined salt caverns as CAS, and have successfully been in operation up to now.

In Huntorf plant, ambient air is compressed in an intercooled process by two separate turbo-compressor units to a maximum pressure of 72 bars. Before it is stored in the CAS, the air is cooled again. The intercooled two-stage compression process limits exergy losses of the diabatic process design without heat storage device, but still more than 25% of the exergy supplied as electrical energy during compression is wasted due to cooling. Identical to the compression process, expansion is carried out in two separate units in series. When the air leaves the cavern in expansion mode, it is first throttled down to a constant pressure of approximately 42 bars before entering the HP combustion chamber.

Compression and expansion trains are connected to each other by the electric machinery. In this way, the electric machine acts as both, electric motor and generator (M/G), and is coupled to the turbomachinery trains via a clutch on each side. Since the HP compressor works at elevated rotational speed, it is coupled by a gear box. For each 1,000 J of electricity output, 800 J of electricity input and 1,600 J of heat input are used. In 2006, after 28 years of operation, the whole expansion train was retrofitted.

Simplified process scheme of the McIntosh plant is shown in Figure 2.7. For each 1,000 J of electricity output, 690 J of electricity input and 1,170 J of heat input are used. Properly assessed, the McIntosh CAES plant in Alabama USA is ~79% efficient and this could be raised to 84% without major difficulty.

Figure 2.8 shows the comparison between McIntosh and Huntorf plants before and after retrofit. Technical parameters comparison are shown in Table 2.2

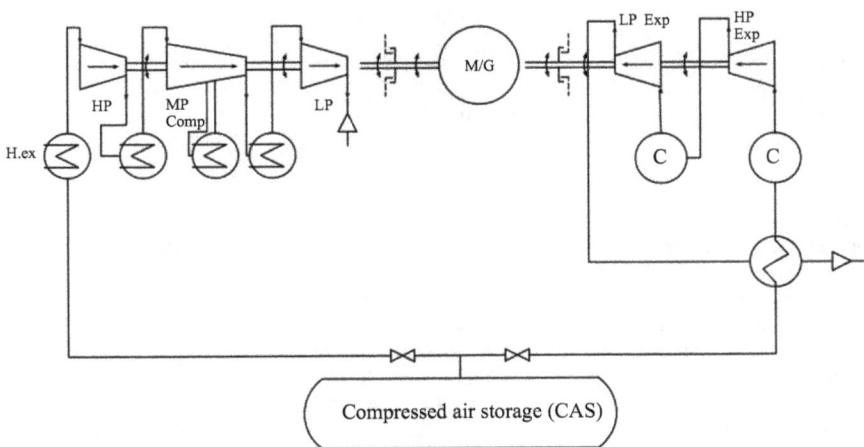

Figure 2.7   Simplified process scheme of the McIntosh plant

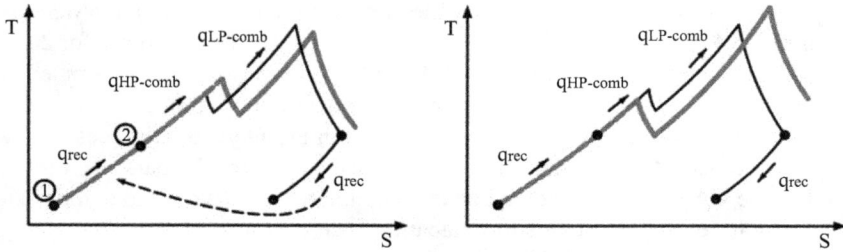

Figure 2.8   T–S diagrams of the expansion process of McIntosh (black line) and Huntorf (grey line) before (left diagram) and after (right diagram) the Huntorf retrofit [4]

Table 2.2   Comparison of technical parameters of operating D-CAES plants [4]

	Huntorf	McIntosch
Plant		
Operating utility	E.ON Kraftwerke	PowerSouth
Cycle efficiency	0.42	0.54
Energy input for 1 kW h_{et} energy output	0.8 kW h_{el}/1.6 kW h_{gas}	0.69 kW h_{el}/1.17 kW h_{gas}
Energy content (related to power output)	642 MWh	2,640 MWh
Planning construction—commissioning	1969–78	1988–91
Compression		
Compressor manufacturer	Sulzer (today MAN Turbo)	Dresser-Rand
Max. el. input power	60 MW	50 MW
Max. air mass flow rate	108 kg/s	Approx. 90 kg/s
Compressor units	2	4
Charging time (at full load)	Approx. 8 h	Approx. 38 h
Storage		
Cavern construction company	KBB	PB-KBB
Cavern pressure range	46–72 bar	46–75 bar
Cavern volume	310,000 m^3	538,000 m^3
Expansion		
Turbine manufacturer	BBC (today Alstom)	Dresser-Rand
Max. el. output power	321 MW	110 MW
Control range (output)	100–321 MW	10–110 MW
Discharging time (at full load)	Approx. 2 h	Approx. 24 h
Start-up time (normal/emergency)	14.8 min	12.7 min
Max. mass flow rate	455 kg/s	154 kg/s
HP turbine inlet	41.3 bar/490 °C	42 bar/583 °C
LP turbine inlet	12.8 bar/945 °C	15 bar/871 °C
Exhaust gas temperature	480 °C	370 °C (before recuperator)

2.4   Combined heat and power dispatch

Combined heat and power (CHP) systems are attracting the increasing attention for their ability to improve the economics and sustainability of the electricity system. Determining how to best operate these systems is difficult because they can consist of many generating units whose operation is governed by complex nonlinear physics. The optimal utilization of multiple CHP systems is a complicated problem that needs powerful methods to solve. The CHP economic dispatch problem is decomposed into two subproblems: the heat dispatch and the power dispatch. The subproblems are connected through the heat-power feasible region constraints of cogeneration units.

2.5   A-CAES systems

CAES systems store the heat of compression and reuse it during the discharging process. As shown in Figure 2.9, this can theoretically be done in two ways.

The simplest way to reuse the temperature-related part of the exergy of the compressed air is to store the hot air itself inside a combined thermal energy and CAS volume. Due to the high temperatures already reached at rather low-pressure ratios, these concepts require highly temperature-resistant storage volumes. The limitations of A-CAES without TES described above lead to the use of a dedicated TES device in most of the A-CAES concepts.

The Goderich A-CAES Facility, located in Goderich, Ontario, Canada, is the world's first commercially contracted advanced-CAES facility.

2.5.1   Thermal storage and cold storage

Applications of TES have been found in the building industry, the automotive industry, and solar energy installation. Recently, more applications are being

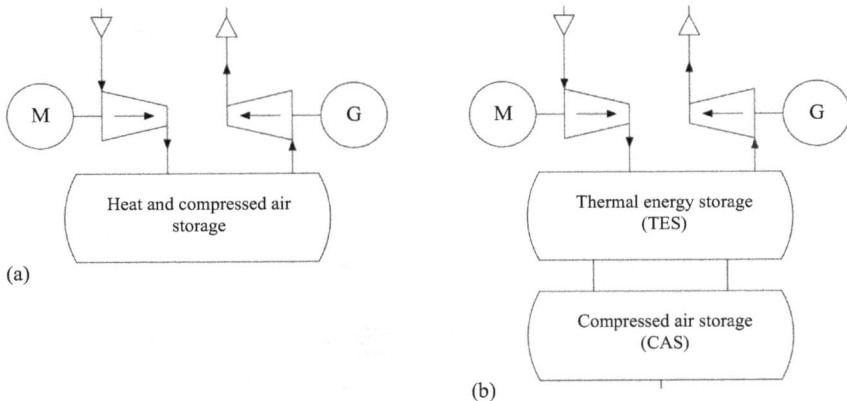

Figure 2.9   Basic concepts of A-CEAS

explored, such as integration of TES with CAES, recycling industrial waste heat, and emissions reduction via replacing nonrenewable energy. TES includes the sensible heat storage (SHS) and the latent heat storage (LHS): SHS uses the heat capacity of the materials for storage/release of heat; LHS uses phase change materials (PCM) to achieve heat storage/release heat.

By removing the temperature-related part of the exergy from the air stream, the cooled pressurized air can be stored in a CAS of any kind. Additionally, much higher final pressures can be addressed and higher energy densities can be reached. Typical final pressures of A-CAES are at least 60 bars. This value is taken as the basis for the following statements. The most important parameter of A-CAES is the chosen storage temperature. It has a direct influence on the system engineering as well as on the operating behavior of the whole storage plant. In contrast, cycle efficiency is hardly dependent on the absolute storage temperature. Corresponding to the considerable differences in terms of system engineering, TES technology, and the resulting operating behavior, three process types can be distinguished:

1. High-temperature processes with storage temperatures above 400 °C.
2. Medium-temperature processes with storage temperatures between 200 °C and 400 °C.
3. Low-temperature processes with storage temperatures below 200 °C.

2.5.2   Advanced A-CAES

The concept of high-temperature A-CAES with single-stage TES was picked up in 2003 by a European research project which resulted in the high-temperature concept called advanced A-CAES (AA-CAES) (see Figure 2.10). As an attractive large-scale clean energy storage technique, AA-CAES can store and generate both electricity and heat, which has great application potentials in integrated electricity and heating systems (IEHSs). AA-CAES is a new technology development direction of conventional CAES (C-CAES). Unlike C-CAES, AA-CAES uses TES to

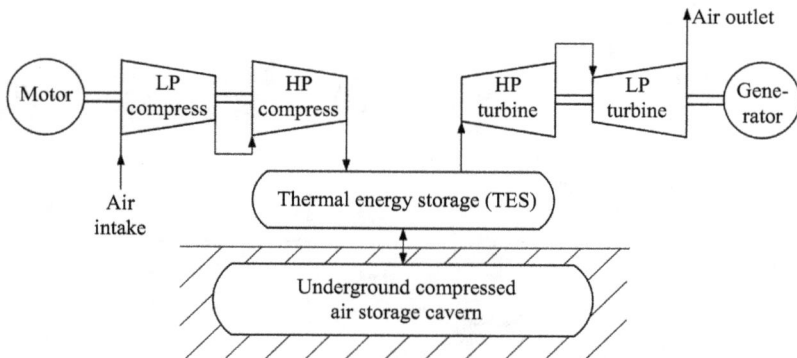

Figure 2.10   One-stage arrangement of an AA-CEAS system

store and reuse compression heat, which makes it, needs no fuel for electric generation, and has higher efficiency than C-CAES. AA-CAES can also easily integrate with EHSs such as solar collectors and electrical heaters, and can use the heat produced by EHSs for heat supply and electric generation. Due to the possession of TES, AA-CAES can not only store and generate electricity, but also store and supply heat. This unique characteristic makes AA-CAES has great application potential in IEHS for wind power accommodation.

In 2012, the *Fraunhofer Institute* proposed a multistage setup using radial compressors and expanders. This design is an A-CAES working at low temperatures and is seen to have efficiencies in the range of 58%–67%. This arrangement is proposed to be superior to the *ADELE* concept due to its capability to integrate in fast start-ups, whilst also having a part-load capability across a wide range. To date, this concept exists only at an academic level and there are no identified manufacturers and suppliers who could provide data through feasibility studies [5].

In 2013, the *ADELE 290 MW* project was proposed as the first adiabatic demonstration land-based plant in Germany, where the compression heat was proposed to be stored having a TES which is separate from the ambient temperature HP air store. The main challenge of this plant is to have high-performance compressors for operating temperatures and pressures up to 600 °C at 100 bars, respectively, with high-efficiency heat storage containers. To date, this project has been delayed [6].

ALACAES of Lugano, Switzerland, is planning the world's first HP AA-CAES plant. This proposed 100-bar plant follows the successful implementation of a 600 kW plant with a capacity of 1 MWh operating at 7-bar plant in 2016 [7]

2.6   Near-isothermal systems

At a smaller scale, serious attempts are being undertaken in the USA toward near-isothermal CAES without fuel consumption. Several start-up companies are developing fuel-free prototypes in the range of hundreds of kilowatts installed power applying reciprocating piston engines to compress and expand air. I-CAES systems try to prevent temperature increase in the compressors during charging and temperature drop during discharging in the expansion devices. All I-CAES concepts known so far are based on piston machinery, since these machines can perform a comparably slow compression or expansion process, which leaves enough time for heat exchange processes inside the machinery itself. For example, the heat exchange can be carried out using additional heat exchange surfaces and a liquid piston. Another way is to spray liquids into the plug room of a common piston machine or the compression of premixed foam.

2.7   Isothermal CEAS systems

I-CAES attempts to achieve near-isothermal compression *in situ*, thus avoiding external heat exchangers to compress/expand air. This yields improved system efficiencies (~70%–80%), provides fuel-free operation, and reduces thermal stress on equipment. Generally, there are several ways of enhancing heat transfer for achieving isothermal compression/expansion. Current I-CAES approaches employ up to a three-stage reversible compressor/expander, with each cylinder realizing up to a 1:30 compression ratio and total pressure up to 400 bars. Three patented I-CAES technologies under development include the following: the injection of liquid (water/oil) into a reciprocating piston cylinder during air compression, or the bubbling of air through the liquid; the separation and collection of that medium into a TES reservoir; and the reinjection of the warm liquid into the cylinder during expansion.

These approaches have been realized in three partially government-funded and now operational pilot-scale I-CAES plants: General Compression (2 MW, 500 MWh 2012); SustainX (1.65 MW, 8 MWh 2013); and Light Sail Energy (2 MW, 8 MWh 2013).

The company SustainX, which spun out of Dartmouth College, had proposed an isothermal process having a water-in-air heat transfer process within pneumatic cylinders. The plant achieves the isothermal characteristics by spraying water in the air which is being compressed in the compression chamber. This design allows heat to be transferred from water to air during expansion or from air to water during compression. Since the same power unit provides both isothermal compression and expansion, the cost of separate compressor and expander subsystems is avoided [8].

Another isothermal design has been suggested by LightSail in California involving spraying water droplets during the compression cycle to absorb the heat of compression and limit the temperature increase in the compression chamber. The warm water is stored and eventually reintroduced during the expansion process as droplets. Apex, a Texas-based company, has proposed the Bethel project being a 317 MW CAES facility with a storage capacity of 30,000 MWh that is enough to energize more than 300,000 homes [9].

2.8 Supercritical CAES

The supercritical CAES (SC-CAES) system is a new type of CAES system which integrates the advantages of both AA-CAES and LAES: environmental protection, high energy density, and high thermal efficiency. Figure 2.11 shows a typical SC-CAES system. The air is compressed to reach to its supercritical state ($P > 37.9$ bar, $T > 132$ K); and then the supercritical compressed air is stored in tanks after a heat exchanger collects the compression heat; the liquid air becomes its gas state and generates power after being pumped to supercritical pressure and heated by the heat exchangers.

There are two ways to perform liquefaction: throttle liquefaction valve and liquefied expander. The cooled effect of air is used by the throttle valve to obtain liquid air in the throttling process, in which the temperature is reduced. Because it is a kind of typical irreversible process, it will not only consume large amounts of energy, but also cause cavitation. The liquid expander is used to replace the throttle

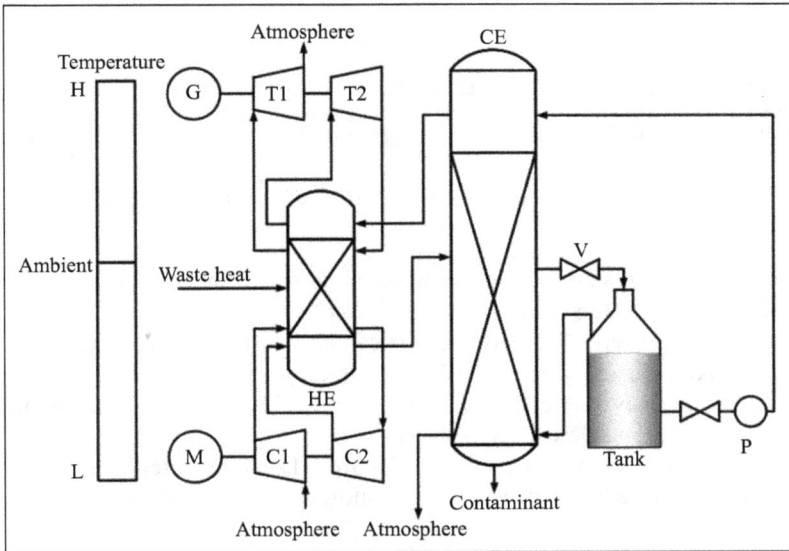

Figure 2.11   The schematic diagram of SC-CAES

valve to achieve the throttling depressurization effect. Volume energy density of compressed air is low and large-scale expensive pressure vessels are necessary for storing a large amount of compressed air. In the LAES and SC-CAES systems, liquid air is stored in insulated tanks at low temperature and normal pressure, thereby achieving higher volume energy density and lower cost for pressure vessels. Compared with A-CAES, the energy density of the LAES and SC-CAES systems is higher because of the liquid state of the air; however, the structures of LAES and SC-CAES systems are more complex and present more energy loss links, thereby contributing to a lower round-trip energy efficiency.

2.9   Underwater CAES

CAES can be done with underground and underwater storage of HP air, and underwater CAES (UWCAES) is very attractive because of constant-pressure characteristic. Underground storage of HP air is especially attractive at large scales. As an alternative to the pump hydro storage (PHS), CAES has become one of the most widely studied energy storage technologies for storing marine renewable energy (MRE).

An obvious distinction between the terrestrial applications and marine applications of CAES is where and how the compressed air is stored. The CAES concepts for MRE storage can also be divided into isochoric and isobaric. As stated before, the isochoric storage is the most common storage method of compressed air, and the compressed air is stored in reservoirs with constant volumes. While in

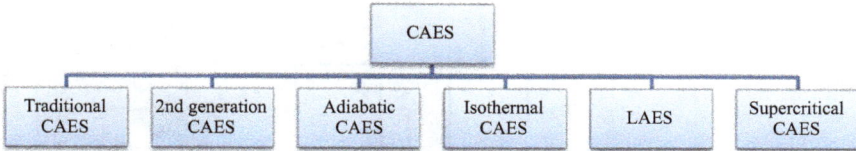

Figure 2.12   Classification of main CAES technologies

an isobaric CAES system, the compressed air is stored at a constant pressure in reservoirs with variable volumes.

The CAES concepts for MRE storage can also be divided into indirect CAES and direct CAES. In an indirect CAES concept, the air is compressed by electricity generated from renewable energies. In a direct CAES concept, the air is directly compressed by renewable energy devices.

All CAES technologies presented in Figure 2.12 can be integrated with MRE. Many novel concepts have been proposed following the basic principles of various CAES technologies.

2.10   Offshore systems

The ability of CAES to compensate for fluctuating renewable energies was mentioned as early as 1976, although without being of major importance at that time. This has now changed dramatically with a significant penetration of intermittent renewable energies such as wind and photovoltaics in the electricity supply system in many countries around the world.

Offshore wind power is especially promising due to high-class wind zones existing near areas of high population density. With the rapid development of MRE technologies, the demand to mitigate the fluctuation of variable generators with energy storage technologies continues to increase.

Offshore CAES (OCAES) is a novel flexible-scale energy storage technology, that is suitable for MRE storage in coastal cities, islands, offshore platforms, and offshore renewable energy farms.

Typical offshore wind farms to date vary in size from 207 MW (*Rodsand II, Denmark*) to 630 MW (*London Array, United Kingdom*). A typical 200 MW CAES plant would need a minimum storage capacity of 10,000 MWh; being the equivalent of a typical 50 h of full plant output of wind turbines, assuming that the wind power density is constant throughout the year. In case of seasonal storage, a minimum of 40,000 MWh would be needed [9].

A number of works have examined the feasibility of using compressed air as an underwater energy storage technology for offshore use. Storage options could be through rigid containers or through energy bags; these being bags fabricated out of specially reinforced fabric.

Rigid containers have been researched at the *University of California (UOC)* using compressed air to pump out water when there is an energy surplus. This compressed air is released when energy is needed to drive the expansion equipment. This research proposes a 230 MW installation with a storage capacity of 10 h and operating pressures of 60 bars in pipework at a depth of 650 m.

At the *Massachusetts Institute of Technology* (*MIT*), the concept of rigid storage is being researched through concrete spheres, where again the flow of water in and out of the reservoir is used in synchronization with the compression and release of the water. This concept introduces also the idea of using the reservoirs as anchorage for energy platforms such as wind turbines. Other concepts of underwater storage using rigid containers have also been investigated using the surrounding bedrock as the thermal reservoir for the compression heat [10].

The *University of Nottingham* has been studying various shapes and cost models for energy bags since 2007. Energy bags studied at this university considered an adiabatic process. The compression heat is contained within a multilayer of the energy bag. The bags have three inner most layers with molten salt in a porous bed of rock fragments to act as a heat transfer medium. The bag has a middle section made up of three layers containing mineral oil as a heat transfer fluid with a capacity to handle temperatures of 250 °C. The three outermost layers are made up mainly of seawater, catering for temperatures of up to 100 °C. The Canadian company *Hydrostor*, together with a number of universities, proposes spherical containers as energy bags for underwater storage, using the hydrostatic pressure of the surrounding water to push the compressed air out of the bags when it is needed to drive the turbomachinery. Their system in Lake Ontario is noted as the world's first grid-connected underwater energy storage facility. In Figure 2.13, different categorizations for offshore CAES are shown [11].

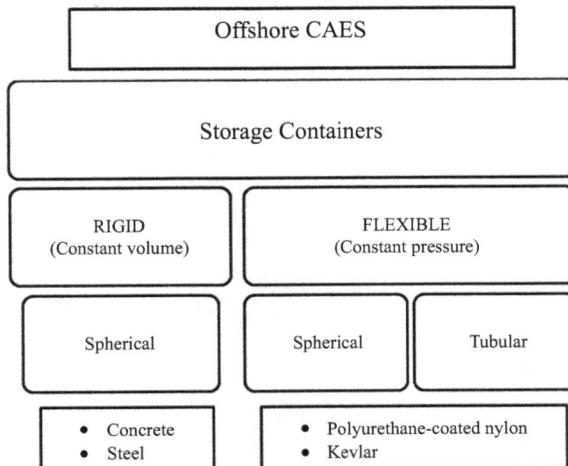

Figure 2.13   Offshore CAES categorizations

2.11   Analysis, simulation, and modeling

Accurate modeling of CAS cavern done by including the mass and energy balance inside the cavern shows that heat transfer plays an important role in determining the behavior of the cavern.

By incorporating accurate heat transfer model, the cavern behavior can be accurately simulated. The mass and energy balance are written over the control volume enclosing the air in the cavern. Figure 2.14 shows the control volume enclosing the cavern storage bed.

Here, ρ is the density of air inside the cavern and m_{in} is the mass flow rate of the incoming air from the compressor, and m_{out} is the mass flow rate of the outgoing air to the turbine and V is the volume of the cavern.

$$\frac{d\rho}{dt} = \frac{\dot{m}_{in} - \dot{m}_{out}}{V} \tag{2.1}$$

Energy balance is written over the control volume enclosing the air in the cavern as

$$\frac{d(MU)}{dt} = \dot{m}_{in} H_{in} - \dot{m}_{out} H_{out} - h_{amb} A_{cavern} (T - T_{amb}) \tag{2.2}$$

Data from Huntorf cavern operation is used to validate model. Comparisons made with the isothermal and adiabatic models are found to inadequately describe the behavior of the cavern as shown in Figure 2.15. Such modeling efforts will be useful in the future. Design of cavern based CAS beds, in terms of making a good estimate of the cavern volume and operating conditions.

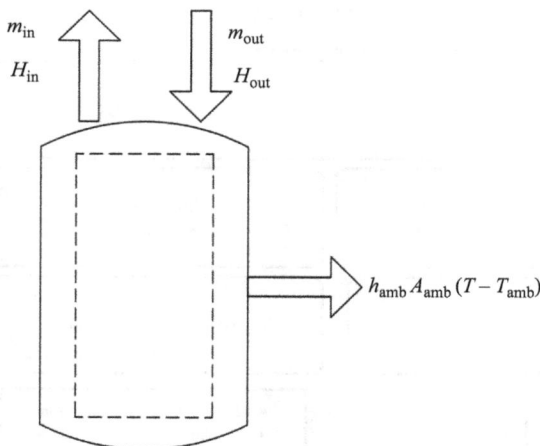

Figure 2.14   Schematic of control volume enclosing the cavern storage bed

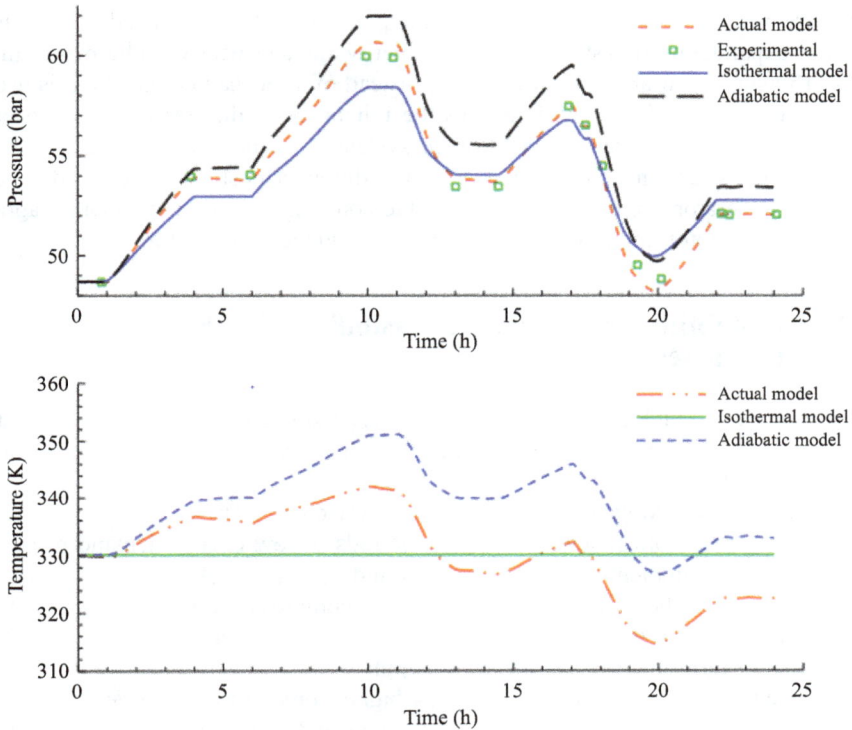

Figure 2.15 Comparison of the cavern characteristics for the current model with the adiabatic and isothermal models [12]

2.12 Power electronics and integration with transmission and grid

Two scenarios of different generation scales and locations can be considered:

1. centralized CAES (grid scale);
2. distributed CAES (small scale).

Both the centralized CAES and distributed CAES plants would help smooth fluctuations. The capacity factor of current offshore wind turbines are typically less than 50%. Therefore, the electrical generator, collection, and transmission system are underutilized. Collection and transmission is a major portion of the balance-of-plant cost for an offshore wind turbine. By storing the energy prior to generation of electricity, the electrical components can be downsized for demand instead of supply.

In CAES for offshore wind turbines, excess energy from the wind turbine is stored locally, prior to electricity generation, as compressed air in a storage pressure vessel. This allows electrical components to be downsized. The compressor/expander used to store and extract energy operates nearly isothermal so that it is efficient. A variable hydraulic drive, instead of a mechanical gearbox, is used for power transmission. This improves the reliability of the transmission system and allows the generator and the storage system to be housed down tower, thus reducing construction and repair costs. In addition, a cost-effective fixed-speed inductor generator can be used instead of the combination of a permanent magnet synchronous motor and power electronics for frequency conversion.

2.13   Reliability and economic feasibility: CAES risk assessment

Cost-effective, scalable, and dispatchable energy storage systems are the key to integrating unpredictable and intermittent green energy, such as wind and solar energy, into the electrical grid.

Generally speaking, CAES plants use lower-cost electricity or surplus electricity generated from renewables during periods of low energy demand to compress and store ambient air in an underground cavern, or aboveground in large storage vessels. When electricity is needed, the compressed air is released, heated up, and expanded in turbines to generate electricity, effectively providing a buffer against short-term interruption of power supply.

Some studies address the risk of storing highly compressed air; large risk factors were categorized in three components: (1) planning and design phase, (2) construction phase, and (3) operation and maintenance phases.

The financial risks imposed from the uncertain parameters are a considerable issue in the optimization problem of renewable-based energy systems.

2.14   Outlook: wind-powered CEAS

Electricity generated from the burning of fossil fuels has many environmental problems, one of which is climate change caused by the large amount of carbon dioxide that is emitted during the fossil fuel combustion process. Dealing with global climate change, therefore, requires developing solutions to reduce the carbon footprint. Increasing the supply of renewable energy would help us replace carbon-intensive energy sources and significantly reduce CO_2 emissions. Wind power is a low carbon energy source and its installed capacity has increased rapidly in the last 10 years. To promote the development of green electricity market, many governments have provided R&D funding, regulations, and financial support to promote the growth of wind generation. Integrating large quantities of wind power into existing electricity grids presents a significant challenge to ensure that electricity supply constantly matches power demand.

The fact is that wind energy is uncertain and variable, and therefore, it is unable to be guaranteed to meet the power demand in the day.

Technically, this variability causes supply imbalance, which will increase the ramping duty and flexibility requirements for coal and gas generators. To maintain the balance between power generation and consumption, many generating units are commonly scheduled for daily load cycle operations, due to system demand variations, resulting in a considerably lower capacity factor for some plants and a higher electricity price at certain times. One of the most promising options, which are capable of managing the fluctuation of wind power, is CAES [13–25].

References

[1] Cipollone R. "Sliding vane rotary compressor technology and energy saving." *Institution of Mechanical Engineers, Journal of Process Mechanical Engineering*, 2016.

[2] Wang J, Lu K, Ma L, Wang J, Dooner M, and Miao SH. "Overview of compressed air energy storage and technology development." *MDPI Journal of Energies*, 2017.

[3] Hea W and Wang J. "Optimal selection of air expansion machine in compressed air energy storage: A review." *Renewable and Sustainable Energy Reviews*, 2018: 77–95.

[4] Budt M, Wolf D, Span R, and Yan J. "A review on compressed air energy storage: Basic principles, past milestones and recent developments." *Applied Energy*, 2016: 250–268.

[5] Budt M and Wolf D. "LTA-CAES—A low-temperature approach to adiabatic compressed air energy storage." *Applied Energy* 125, 2014: 158–164.

[6] Power R. "ADELE—Adiabatic compressed air energy storage for electricity supply." RWE PowerAG, 2010.

[7] Zavattoni SA, Geissbühler L, Zanganeh G, Haselbacher A, Steinfeld A, and Barbato MC. "CFD modeling and experimental validation of the TES unit integrated into the world's first underground AA-CAES pilot plant." *AIP Conference Proceeding*. AIP Publishing, 2019.

[8] Bollinger B. *Technology Performance Report: SustainX smart grid program.* Technology Performance Report, SustainX, 2015.

[9] Vella P, Sant T, and Farrugia RN. "A review of offshore-based compressed air energy storage options for renewable energy technologies." *9th European Seminar OWEMES*, 2017.

[10] Pimm AJ, Garvey SD, and Jong MD. "Design and testing of energy bags for underwater compressed air energy storage." *Energy*, 2014: 496–508.

[11] Jong MD. "Commercial grid scaling of energy bags for underwater compressed air energy." *Proceedings of 2014 Offshore Energy & Storage Symposium*. Ontario, Canada, 2014.

[12] Raju M and Khaitan SK. "Modeling and simulation of compressed air storage in caverns: A case study." *Applied Energy*, 2012: 474–481.

[13] Balje OE. "A study on design criteria and matching of turbomachines: Part A. Similarity relatives and design criteria of turbines." *Trans. ASME Series Journal of Engineering for Power*, 1962: 83–102.

[14] Carriveau R, Ebrahimi M, Ting DS-K, and McGillis, A. "Transient thermodynamic modeling of an underwater compressed air energy storage plant: Conventional versus advanced exergy analysis." *Sustainable Energy Technologies and Assessments*, 2019: 146–154.

[15] Jafarizadeh H, Soltani M, and Nathwani J. "Assessment of the Huntorf compressed air energy storage plant performance under enhanced modifications." *Energy Conversion and Management*, 2020.

[16] Japikse D and Di Bella AF. "An analysis of an advanced compressed air energy system (CAES) using turbomachinery for energy storage and recovery and for continuous on-site power augmentation as an air Brayton cycle." *Mechanics and Mechanical Engineering*, 2018: 479–493.

[17] Liu JL and Wang JH. "A comparative research of two adiabatic compressed air energy storage system." *Energy Conversion and Management*, 2016: 566–578.

[18] Luo X, Wang J, Dooner M, and Clarke J. "Overview of current development in electrical energy storage technologies and the application potential in power system operation." *Applied Energy*, 2015: 511–536.

[19] Saadat M and Li P. "Modeling and control of a novel compressed air energy storage system for offshore wind turbine." *American Control Conference*, Montreal, 2012, 3032–3037.

[20] Vasel-Be-Hagh AR, Carriveau R, and Ting D. "Underwater compressed air energy storage improved through Vortex Hydro Energy." *Sustainable Energy Technologies and Assessments*, 2014: 1–5.

[21] Venkataramani G, Parankusam P, Ramalingam V, and Wang J. "A review on compressed air energy storage—A pathway for smart grid and polygeneration." *Renewable and Sustainable Energy Reviews*, 2016: 895–907.

[22] Wang ZH, Carriveau R, Ting DS-K, Xiong W, and Wang Z. "A review of marine renewable energy storage." *International Journal of Energy Research*, 2019: 1–43.

[23] Wang ZH, Xiong W, Ting DS-K, Carriveau R, and Wang Z. "Conventional and advanced exergy analyses of an underwater compressed air energy storage system." *Applied Energy*, 2016: 810–822.

[24] Wang ZH, Xiong W, Ting DS-K, Ceriveau R., and Wang Z. "Comparison of underwater and underground CAES systems for integrating floating offshore wind farms." *Journal of Energy Storage*, 2017.

[25] Yaowang L, Shihong M, Binxin Y, Ji H, Shixu ZH, Jihong W, and Xing L. "Combined heat and power dispatch considering advanced adiabatic compressed air energy storage for wind power accommodation." *Energy Conversion and Management*, 2019.

Chapter 3
Isothermal compressed air energy storage

Xinjing Zhang¹, Haisheng Chen¹, Yujie Xu¹,
Xuezhi Zhou¹ and Huan Guo¹

Isothermal compressed air energy storage (I-CAES) technology is considered as one of the advanced compressed air energy storage technologies with competitive performance. I-CAES has merits of relatively high round-trip efficiency and energy density compared to many other compressed air energy storage (CAES) systems. The main challenge is to realize high-efficiency heat transfer for charging and discharging in order to keep the air temperature almost constant, thus, to achieve the isothermal or near-isothermal compression and expansion. In this chapter, the general concept of I-CAES was introduced, and its thermodynamic cycle was illustrated. The research progress of I-CAES was reviewed. Both I-CAES system configuration and performance, and key components performance were reviewed. A new isothermal expander was further designed for I-CAES. The specific reciprocating expander with a high-pressure ratio was developed and its adiabatic expansion characteristics were measured by the authors' group. We further proposed a quasi-isothermal expansion process using water injection into the expander cylinder. Modelling was developed and validated by the experimental results of the adiabatic expander, which was also extended to simulate the quasi-isothermal process by introducing water–air direct heat transfer equations. Simulation results showed that when spraying tiny water droplets into the cylinder, the specific work generated was improved compared with that of the adiabatic expansion under the same air mass flow rate. While the temperature difference between inlet and outlet was decreased substantially, and the cylinder size was more compact. The influence of water/air mass flow rate ratio and the inlet temperature on the expander performance was also studied.

Keywords: Isothermal compressed air energy storage; working principle; isothermal expander; experiment and simulation; pressure ratio; specific work

¹Institute of Engineering Thermophysics, Chinese Academy of Sciences, Beijing, China

Nomenclature

Symbols	Concepts	Units
A	Open area of the inlet/outlet valve	m^2
BDC	Bottom dead centre	
c_p	Specific heat at constant pressure	kJ/(kg K)
c_v	Specific heat at constant volume	kJ/(kg K)
CAES	Compressed air energy storage	
C_D	Drag coefficient	
d	Water droplet diameter	m
D	Cylinder diameter	m
g	Acceleration	m/s^2
h	Specific enthalpy	kJ/kg
I-CAES	Isothermal compressed air energy storage	
m_a	Mass of air	kg/s
m_w	Mass of water	kg/s
Nu	Nusselt number	
P	Pressure	Pa
Pr	Prandtl number	
Q	Heat exchange through the cylinder walls	kJ
Q_c	Transferred heat between water and air	kJ
S	Stroke distance	m
S_w	Water droplets surface area	m^2
T	Temperature	K
TDC	Top dead centre	
t	Time	s
u	Specific internal energy	kJ/kg
V	Volume	m^3
v	Velocity	m/s
w	Velocity	m/s
W	Mechanical energy	kJ
W_{adia}	Work generation of adiabatic expansion	kJ
W_{s_adia}	Ideal work generation of adiabatic expansion	kJ
W_{n_iso}	Work generation of near-isothermal expansion	kJ
W_{iso}	Ideal work generation of isothermal expansion	kJ
W_{pump}	Work consumption of water pump	kJ

Greek symbols

α	Heat transfer coefficient	
η_{adia}	Isentropic efficiency	
η_{isoth}	Isothermal efficiency	
η_{isoth2}	Isothermal efficiency considering pump power consumption	
κ	Specific heat ratio	
λ_e	Ratio of connected rod length to crank radius	
λ	Thermal conductivity	W/(m K)
μ	Gas flow coefficient	
μ_a	Dynamic viscosity of the air	Pa s
ρ	Density	kg/m^3
φ	Crank angle	rad

(*Continued*)

Symbols	Concepts	Units
ω	Angular speed	rad/s
ψ	Flow function	
Subscripts		
a	*air*	
A	Outlet for air phase, accumulated water on the piston for water phase	
E	Inlet	
I	Before the inlet/outlet valve	
II	After the inlet/outlet valve	
out	Outlet	
pump	Pump	
w	Water	

3.1 Introduction

Energy storage can transform intermittent renewables into high-value energy that significantly contribute to climate change mitigation [1]. Energy storage can manage both electricity deficits and surpluses [2], and reduce curtailment and CO_2 emission potentially by being integrated into renewables [3]. Arbabzadeh *et al.* [3] concluded that adding 60 GW renewables to California can reduce 72% CO_2 relative to a zero-renewable case with close to one-third of renewables being curtailed. This case was operated as a baseload with a 7.0 GW minimum-dispatchability. Moreover, energy storage deployment (e.g. CAES) with the renewables yields higher CO_2-emission reductions to be 90%, and reduces the curtailment to be as little as 9%. CAES, and is currently the only other commercially mature technology for grid-connected application [4].

Many types of CAES have been studied and developed, including conventional CAES, advanced-adiabatic CAES, liquid air energy storage, isothermal CAES, and so forth [5]. The I-CAES has merits of relatively high round-trip efficiency and energy density compared to other CAES systems. In the ideal I-CAES process, the temperature during compression is kept constant while related heat is released. The power required to run the compressor is correspondingly lower than that required to run an adiabatic compressor with the same pressure ratio and mass flow rate. During expansion, related heat is supplied continuously to ensure expansion at a constant temperature. Thus, the electrical power used to run the compressor during charging can be completely recovered during discharging. The ideal cycle efficiency of I-CAES systems can be as high as 100% [5,6].

3.2 I-CAES working principle

I-CAES refers that the air temperature is kept constant or nearly constant during charging and discharging, and it is stored near ambient temperatures [7]. The temperature rise of the compressed air is assumed to rise in quasi-equilibrium steps where heat is transferred almost instantaneously, thereby preventing the compression-process temperature rise and expansion-process temperature drop [7]. Thus, the system is named isothermal CAES or near-isothermal CAES. Generally, most gas power cycles perform work by expanding adiabatically and produce less specific work than their isothermal counterparts, which run at constant temperature with the same inlet temperature, e.g. room temperature. Isothermal expansion leads to a high-pressure ratio and high power density, improving specific work generation. Meanwhile, lower inlet/outlet air temperature differences result in better performance in low-grade heat applications [8]. If heat is continuously transferred to the working fluid during the expansion process, this results in isothermal expansion or, more accurately, quasi-isothermal expansion, which may be somewhere between the adiabatic and ideal isothermal processes [9]. The major challenge is how to enhance the heat transfer between the compressed air and another fluid or a device.

There are generally three methods to keep the temperature constant: secondary fluid heat transfer (direct heat transfer); compressor/expander surface heat transfer (indirect heat transfer); and multi-stage compressor/expander with intercoolers/interheaters (indirect heat transfer) [7–11]. The operation principle can be illustrated in Figure 3.1. (a) For the secondary fluid heat transfer method, the ambient air is mixed with another high thermal capacity fluid while compressed. The compressed heat is directly transferred to the secondary fluid, e.g. spayed water droplets. Thus, the air temperature increase is kept almost constant. The water is separated after the compressor through a separator, and the compressed air is stored in the storage vessel and the fluid is stored in the high-temperature thermal storage

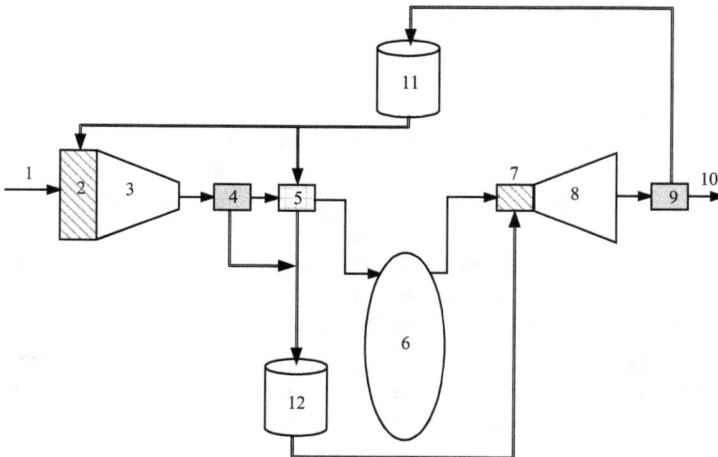

1. Ambient air, 2. mixer, 3. compressor, 4. separator, 5. heat exchanger, 6. storage vessel, 7. mixer, 8. expander, 9. separator, 10. exhaust air, 11. low-temperature TES, and 12. high-temperature thermal energy storage.

Figure 3.1   Schematic diagram of the I-CAES system

tank (TES). It is similar in the discharging process. The compressed air from the vessel is released into the expander, and it is mixed with high thermal capacity fluid while expanding to slow down the temperature decrease. The fluid from the high-temperature TES is sprayed into the expander, and then separated afterwards to enter the low-temperature TES. The heat can be transferred in a quick manner through this method, and the pressure ratio of one stage can be very big since the temperature is almost constant [8]. (b) For the surface heat transfer, the heat from or to the compressed air is transferred through a surface. The high thermal capacity fluid and compressed air are not mixed [12]. The cylinder surface should be large and certain geometries used to enlarge the heat transfer area, such as tubes and fins. Suggestions have also been made to decrease the rotation speed of the expander in order to increase the heat transfer time for a certain amount of fluid [11–13]. However, practice shows that the heating/cooling effect of this kind of external heating/cooling method is limited [14]. As shown in Figure 3.1, the fluid from low-temperature TES flows through the compressor cylinder to cool down the compressed air. It still needs a heat exchanger after the compressor (before the expander) for the limited heat transfer effect through a surface. More stages of compressor/expander might be required to reach a certain air pressure. (c) For multi-stage compressor/expander with intercoolers/interheaters, the compression/expansion processes are divided in to many stages. It might be an adiabatic compression/expansion for each stage. However, the pressure ratio of each stage is small leading to a limited temperature increase. The whole compression/expansion processes is a polytropic procedure, which can be regarded as a near-isothermal process. As shown in Figure 3.1, the air enters compressor directly and the compression heat is absorbed through the heat exchanger after the every compressor stage. The compression heat is stored and reused through another heat exchanger before the expander. It is noted that more stages (e.g. six stages) are operated as compressor chain/expander chain. Moreover, another I-CAES concept of liquid piston is proposed and studied [6,15,16]. A liquid piston-based compressor/expander is similar to the secondary fluid method for direct heat transfer. The difference is that a liquid piston is developed to replace the conventional metal piston to further enhance the heat transfer, and decrease the friction and leakage between piston and the cylinder.

Thermodynamic diagrams are presented in Figures 3.2 and 3.3. Both pressure–volume (P–V) and temperature–entropy (T–S) diagrams of isothermal and adiabatic processes are illustrated. It is noted that the first two of the above-mentioned three methods to achieve isothermal process were regarded as the same with the respect to dissipate the compression heat/add enough heat in expansion, and control the temperature constant ideally. The losses were neglected as the efficiencies of all components in the cycle were assumed to be 100%. The compression, expansion, and storage pressures are set to be the same. The compression and expansion are estimated one stage each for I-CAES, and one stage each and three stages each for adiabatic CAES, respectively. TL1 and TL2 are two constant-temperature process lines. PL1 and PL2 are two constant-pressure process lines. For the one-stage I-CAES system, during charging, the air is compressed isothermally from point 0 to point 6, and then stored. During discharging, the air pressure is first decreased to a

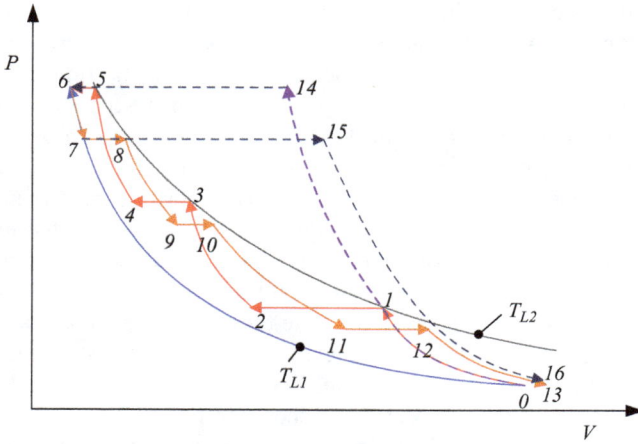

Figure 3.2   Pressure–volume diagram illustrating isothermal and adiabatic processes

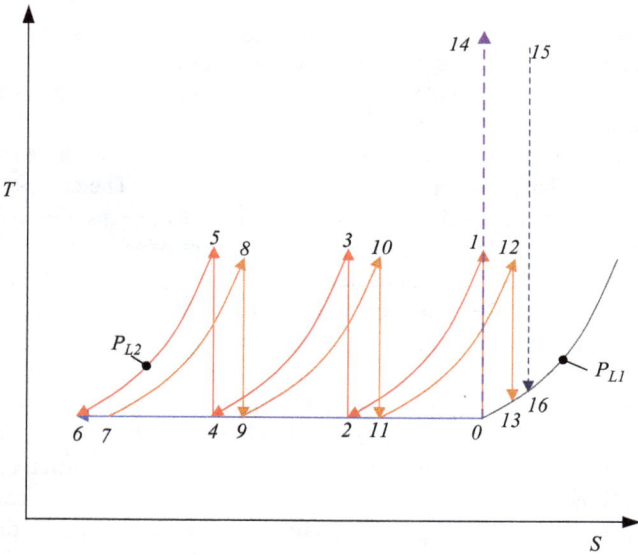

Figure 3.3   Temperature–enthalpy diagram illustrating isothermal and adiabatic processes

certain value and expanded isothermally (6-7-0). For the one-stage adiabatic CAES, during charging, the air is compressed adiabatically from point 0 to point 14, then cooled down and stored at point 6. During discharging, the air is released to a certain pressure and heated, then expanded adiabatically (6-7-15-16). For three-stage adiabatic CAES system with intercoolers and interheaters, pressure ratio of each stage is assumed to be the same, and the outlet temperature of each stage is almost the same as well. The compression is of the curve 0-1-2-3-4-5-6, and expansion 6-7-8-9-10-11-12-13. These two curves are closer to the isothermal line (0-6) if more stages added to both compression and expansion processes.

3.3   Research progress of I-CAES

3.3.1   *Systematic configurations and performance*

Oak Ridge National Laboratory pointed out that compressing and expanding a gas nearly isothermally allows efficiency losses due to temperature deviations to be minimized or eliminated, which can in turn prevent heat transfer loss leading to improved efficiency [17]. The laboratory proposed the Ground-Level Integrated Diverse Energy Storage (GLIDES) system which stores energy via gas compression and expansion (Figure 3.4). The GLIDES concept utilized higher efficiency hydraulic machines for expansion and compression, which is a liquid piston forming a leak-free seal [17]. The system indicated efficiency is then calculated by dividing the work output by the work input. An experimental system using condensable gas of R134a was built to investigate energy storage potential and compression/expansion characteristics. The experimental results showed a round-trip efficiency of 95.8% was achievable [17].

An open-type I-CAES(OI-CAES) system integrated with spray cooling is proposed as shown in Figure 3.5 [18]. It is mainly composed of a compressed air storage (CAS) tank, two working cylinders for air compression/expansion, a reversible hydraulic pump/turbine for water flow, a motor/generator to store/generate electricity, and two pumps to spray water. The energy storage medium is air and the power generation medium is water. During charge /discharge period, the two cylinders operate one by one with all valves on or off to generate compressed air/electricity. The air pressure is 10 MPa, and the round-trip efficiency can reach 76% [18].

Perry Li *et al.* [19] proposed an I-CAES system particularly suited for offshore wind turbines (Figure 3.6). The system uses an open accumulator architecture and a near-isothermal compressor/expander without the use of hydrocarbon fuel as in conventional CAES systems. The maximum pressure is 20 MPa. A concept of

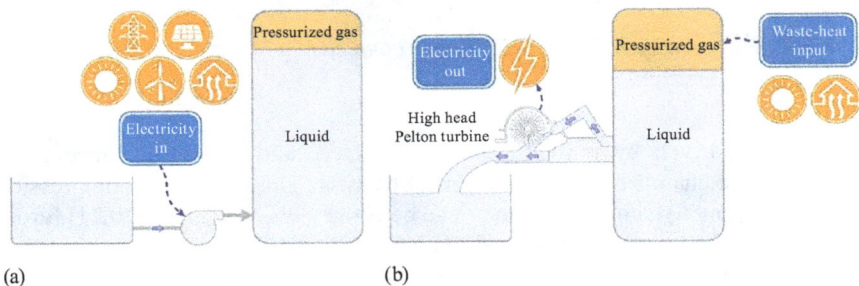

(a)   (b)

Figure 3.4   GLIDES layout during (a) charging and (b) discharging [17]

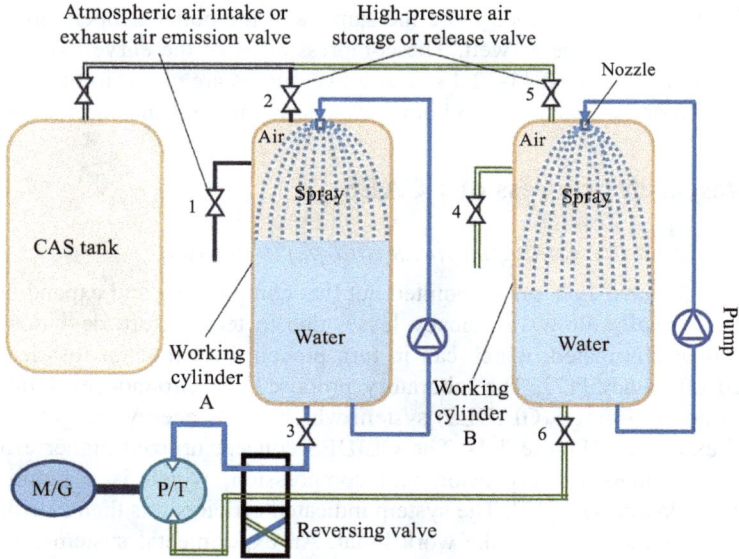

Figure 3.5   Scheme diagram of OI-CAES with spray cooling/heating [18]

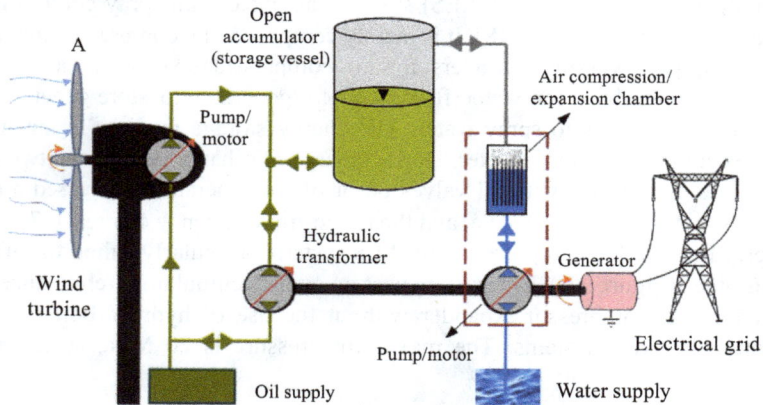

Figure 3.6   Open accumulator CAES system coupled to a hydraulic wind turbine [19]

liquid piston driven by pump/motor was utilized, and the porous media was installed in the chamber for enhancing heat transfer. The isothermal compression/ expansion thermodynamic efficiencies can be above 90%. Qin *et al.* [20,21] further developed the concept, and proposed the I-CAES with hydraulic power transmission (HPT) to the offshore wind farm. Take the NREL (National Renewable Energy Laboratory) 5 MW reference wind turbine at the site of Virginia offshore as an example; it was estimated savings of 21.6% with I-CAES + HPT for an offshore wind farm.

SustainX, Inc. developed an I-CAES demonstration [22]. A 1.5 MW I-CAES prototype was designed and tested. The processes were based upon the compression and expansion of a foam–air mixture to facilitate fast heat transfer and maintain constant temperature throughout. The storage pressure range is up to 20 MPa. The air expansion with water spray maintained a 15 K difference with ambient temperature, and isothermal efficiency achieves 95%. The system round-trip efficiency is 54%.

3.3.2   *Key components analysis*

The compressor and expander are the pivotal components of the CAES system [23]. Generally, most gas power cycles are performed by compressing/expanding adiabatically, they and consume more specific work/produce less specific work than their isothermal counterparts, which run at constant temperature. The operation pressure could be of 20–35 MPa, and currently 1.5 MW I-CAES pilot was developed [22].

In compression, Li *et al.* [14] conducted experimental study of a reciprocating compressor. Tiny water droplets were ejected into the compressor cylinder through different nozzles. The atomization quality and droplet diameters were considered and examined in the experiment. The cylinder outlet temperature was decreased over 50 K compared with near-adiabatic compression. The polytropic component was reduced to 1.161 [14]. Zhang *et al.* [24] investigated the way to increase heat transfer surface area by inserting porous media. An experimental investigation on heat transfer with porous inserts during compression for a pressure ratio of 10 and during expansion for a pressure ratio of 6 was performed. The configuration of inserts was the following: three interrupted ABS inserts with plate spacing of 2.5, 5, and 10 mm and two aluminium foam inserts sized with 10 and 40 pores per inch. In compression, the porous inserts increased power density by 39-fold at 95% efficiency and enhanced efficiency by 18% at 100 kW/m^3 power density; in expansion, power density was increased threefold at 89% efficiency, and efficiency was increased by 7% at 150 kW/m^3 power density [24,25]. Cai *et al.* [26] proposed a contact heat transfer method which is used to cool the compressed air by injecting micron-sized (10–100 μm) water spray into the compressed air. A transient temperature measurement method is developed to investigate the heat transfer behaviour between air and water. The measurements showed the spray-air heat transfer rate to lie in the range of 10–120 W/m^2 K^1. The achieved compression efficiency is improved from 86.7% (adiabatic) to 92.4%. Lee *et al.* [27] investigated an isothermal compressor for a supercritical CO_2 Brayton cycle. The isothermal compressor when CO_2 is the working fluid has been shown to reduce compression work up to 50% in comparison to the conventional compressor when operated near but above the critical point. The cycle performance was improved using this isothermal compressor [27]. Two modelling strategies for isothermal reciprocating compressors were implemented in reference [28]; namely, a simple zero-dimensional (0D) differential equation model alongside a two-dimensional (2D) computational fluid dynamics (CFD) model. The comparison of results between 0D and 2D models shows a very good match for both speeds. It is shown that lower operational speed can result in greater heat transfer from gas, demonstrating a near-isothermal situation. The parametric study carried out using 0D model showed that reducing the dead volume, cooling the cylinder, and reducing the compressor speed can reduce the specific work [28]. Thermal stress of a low-flow long-stroke quasi-isothermal compressor was analysed [29]. It is concluded that intense cooling of the stage has positive effect on the thermal stress state of the compressor stage parts and provides, in comparison with the soft-cooled stage, increasing of the structure reliability under high discharge pressures. Moreover, intense cooling can be considered as a factor, which leads to the decrease of the structure metal content if comparing with uncooled structures.

Liquid piston compression is an exciting concept that can significantly improve the efficiency of compressing or expanding a gas. Through a simplified model, this concept was demonstrated to improve the efficiency of compression from 70% to over 83% [30]. References [18] and [19] investigated liquid compression and expansion for I-CAES system, and efficiencies above 90% could be reached. Isothermal expansion leads to a high pressure ratio and high power density, improving specific work generation. Meanwhile, lower inlet/outlet air temperature differences result in better performance in low-grade heat applications [8]. Cicconardi *et al.* proposed using many super-heaters to make the expansion gradually approach isothermal conditions in a steam power plant and found that the thermal efficiency can be improved from 38.5% to 49.2% [31].

Kim *et al.* [32] indicated that multi-stage compression with intercooling and multi-stage expansion with reheating could also transform the adiabatic process into a near-isothermal process. Thus, the exergy losses due to heat transfer were decreased by minimizing the temperature differences between heat exchangers. System efficiency was as high as 71.6% [32]. Woodland *et al.* [33] carried out theoretical study of the organic Rankine cycle (ORC) based on a liquid flooding expander. For ammonia, an improvement of 20% in thermal efficiency could be reached [33]. Underwater/ocean CAES systems have been investigated and can reach near-isothermal processes with a liquid-piston-based compression cycle to increase the air storage pressure and mitigate thermal loss from high-temperature heat exchange and increase the air storage pressure [34,35]. Both theoretical and experimental studies were conducted. The polytropic index for the compression process was approximately 1.25, and this number decreased as the stroke time was increased [34,35]. An isothermal expander for zero emission automobiles was proposed and examined [11,13]. Ambient heat exchangers were used to power an engine that was configured to maximize heat transfer during the expansion stroke. It can realize specific energies in the range of 300–450 kJ/kg. Lemort *et al.* [36,37] proposed a flooded scroll expander with R245fa working fluid to approach an isothermal process; the best overall isentropic efficiency of the expander was 66% [36,37].

Humid air turbine (HAT) is realized by injecting water into the gas turbine inlet air to achieve a better flexibility and lower NO_x emission. For a micro-gas turbine (rated electric power of 80 kW, inlet air mass flow of 0.73 kg/s), the specific output power was increased by 45%, and a maximum NO_x reduction of about 33% with injection water of 0.007 kg/kg dry air [38]. Siemens [39,40] developed wet compression as a reliable and proven method of injecting water into the gas turbine inlet. Wet compression was designed to increase the power output of the gas turbine by reducing compressor inlet temperatures, intercooling the air mass flow within the compressor and hence an increasing mass flow throughout the turbine. It can also offer attractive financial payback options. A Siemens E-class gas turbine in Dubai was using water injection into the compressor inlet. The power output is increased by 15%, efficiency increased by 0.3%, and NO_x emission decreased by 15% [40].

In this study, we developed a new isothermal expander for I-CAES system. Both theoretical and experimental studies were conducted. The performance of the isothermal expander was presented and analysed.

3.4   System description

A small-scale I-CAES was studied with key components of a reciprocating isothermal expander. Our research group has previously developed and measured a novel small expander [41]. This was a reciprocating expander with only single-inlet valve for a small-scale CAES prototype system. The outlet was near the bottom dead centre (BDC), which was controlled by the piston movement. Fewer movement parts were configured compared with other reciprocating expanders. The expander's internal flow characteristics and adiabatic performance were investigated, which was reported in reference [41]. Moreover, the working fluid of this type of reciprocating expander can be various such as compressed air, water steam, and organic fluids [42–46]. The compressed air can be used to drive the expander for either electricity generation or mechanical power.

A three-stage, single-valve reciprocating expander, as shown in Figure 3.7, was designed and manufactured. It is a small-scale expander of around 10 kW. Based on the manufactured adiabatic expander, we proposed to upgrade it to be an isothermal expander. The isothermal configuration, as shown in Figure 3.8, is that an annular spraying nozzle is installed along with the intake valve. The nozzle injects atomized water at a constant mass flow rate. During the air intake process, the compressed air flows in, mixes with the water droplets, and expands as the piston moves downwards. Meanwhile, the water accumulates on the piston head at a constant flow rate. When the piston comes near the BDC, the expanded air as well as the accumulated water is exhausted into the buffer tank through the outlet orifices and cylinder air pressure further decreases. After that, the remaining air is compressed as the piston moves back to the top dead centre (TDC), and hence, another cycle begins.

The study was first to construct and adiabatic modelling to simulate the expander. The model developed was also validated by experimental results. The results showed that the expander operated with high pressure and relatively high aerodynamic efficiency. Building off of this research, in the current study, we

Figure 3.7   *A three-stage reciprocating expander*

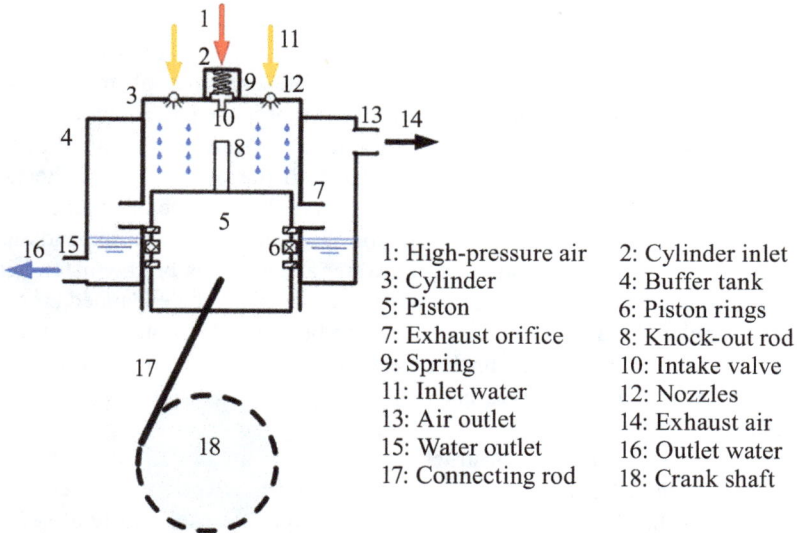

1: High-pressure air	2: Cylinder inlet
3: Cylinder	4: Buffer tank
5: Piston	6: Piston rings
7: Exhaust orifice	8: Knock-out rod
9: Spring	10: Intake valve
11: Inlet water	12: Nozzles
13: Air outlet	14: Exhaust air
15: Water outlet	16: Outlet water
17: Connecting rod	18: Crank shaft

Figure 3.8   *Schematic diagram of the single-valve reciprocating expander*

proposed a quasi-isothermal expander. The near-isothermal process was achieved by spraying water droplets into the cylinder, where direct-contact heat exchange can occur between the gas and the liquid. Compared with previous research, this concept used direct heat transfer and was different from liquid pistons in that the operation pressure could be very much higher than that of scroll expanders due to the self-sealing configuration of the expander. It is also highly applicable since the reciprocating expander had been developed and tested. The multi-phase heat exchange process was modelled and simulated. The expander performance was predicted and a comparison study between adiabatic expansion and dynamic air pressure and temperature inside the cylinder was performed. A parameter-sensitive analysis was also conducted in terms of water/air mass flow rate ratio and inlet temperature.

3.5   Methodology

3.5.1   Experiment description

The test bench of the fabricated three-stage expander, as shown in Figure 3.9, includes the expander, compressor and storage tank, inter-stage heater, pressure/temperature/flow rate sensors (P, T, FR), and the apparatus for controlling speed and inlet pressure. The expander was connected to a motor through a speed and torque transducer. A converter was introduced and connected to the motor and a resistor. The converter was used to control the speed of the motor and also transfer generated electricity to the resistor for consumption. Air pressure and temperature were measured before and after each stage, and a high-frequency pressure trans-ducer was also installed in each cylinder to measure the internal pressure of each stage. The test inlet/outlet pressure of each stage was as follows: 7.03/1.82 MPa for

Figure 3.9   The expander experimental set-up

first stage, 1.82/0.51 MPa for second stage, and 0.51/0.10 MPa for third stage. The measured results of the first stage were analysed, and were utilized to validate the modelling of adiabatic expansion.

3.5.2   Simulation and modelling

3.5.2.1   Modelling of adiabatic expansion process

The modelling process mainly contained two sections: (a) the compressed air expansion process and (b) the process of water droplet spraying/direct heat exchange with air. The air was assumed to act as an ideal gas, and parameters within the cylinder were assumed to be uniform.

The mass balance of air in the cylinder of each cycle was calculated as follows:

$$\frac{dm_a}{d\varphi} = \frac{dm_{a,E}}{d\varphi} - \frac{dm_{a,A}}{d\varphi} \tag{3.1}$$

Once the pressure difference before and after the inlet/outlet valve and the valve open area were determined, the transient mass flow rate could be calculated by the following equation:

$$\frac{dm_{a,j}}{d\varphi} = \frac{1}{\omega} m_{a,j}^{\bullet} = \frac{1}{\omega} \mu A \left(\sqrt{2P_I \rho_I} \right) \psi \tag{3.2}$$

The air temperature in the cylinder changes along with the crank angle, and the equation is as follows:

$$\frac{dT_a}{d\varphi} = \frac{1}{m_a c_v} \left[\frac{dQ}{d\varphi} + \frac{dQ_C}{d\varphi} - \frac{dW}{d\varphi} + h_{a,E} \frac{dm_{a,E}}{d\varphi} - h_{a,A} \frac{dm_{a,A}}{d\varphi} - u_a \frac{dm_a}{d\varphi} \right] \tag{3.3}$$

The mechanical power was calculated as follows:

$$\frac{dW}{d\varphi} = P_a \frac{dV_a}{d\varphi} \tag{3.4}$$

The volume change with the crank angle was calculated as

$$\frac{dV_a}{d\varphi} = \frac{\pi}{8} D^2 S \left[\sin \varphi + \frac{\lambda_e \sin \varphi \cos \varphi}{\sqrt{1 - \lambda_e^2 \sin^2 \varphi}} \right] \tag{3.5}$$

The transient pressure inside the cylinder was calculated by the following equation:

$$P_a = \frac{m_a R T_a}{V_a} \tag{3.6}$$

The thermodynamic properties of the ith stage cylinder can be simulated using the above equations.

3.5.2.2 Modelling of water droplets and heat exchange

Water was utilized to mix with compressed air for direct heat exchange because of its high heat capacity and affordable cost. Since evaporation of the liquid was extremely small, liquid–gas mass diffusion has been ignored in this study.

It was assumed that a single droplet falls at constant terminal velocity. Therefore, the drag and the gravity forces on each droplet are balanced, and the terminal velocity can be calculated with the following equation [9]:

$$v_w = \sqrt{\frac{4d \times \rho_w \times g}{3\rho_a \times C_D}} \tag{3.7}$$

There are many correlations of drag coefficients [47,48]. The following empirical correlations were cited in the calculation:

$$
\begin{array}{ll}
C_D = 24/Re_d; & Re_d \le 0.2 \\
C_D = 18.5/Re_d^{0.6}; & 0.2 < Re_d \le 500 \\
C_D = 0.44; & 500 < Re_d < 10^5
\end{array} \tag{3.8}
$$

$$Re_d = \frac{d \times \rho_a \times |w|}{\mu_a} \tag{3.9}$$

The water temperature in the cylinder changes along with the crank angle, and the equation is

$$\frac{dT_w}{d\varphi} = \frac{1}{m_w c_{p,w}} \left[h_{w,E} \frac{dm_{w,E}}{d\varphi} - h_{w,A} \frac{dm_{w,E}}{d\varphi} + \frac{d(P_w V_w)}{d\varphi} - \frac{dQ_C}{d\varphi} - u_w \frac{dm_w}{d\varphi} \right] \tag{3.10}$$

The heat exchange between the air and water droplets can be calculated as follows [49]:

$$\frac{dQ_C}{d\varphi} = \frac{\alpha S_w (T_w - T_a)}{\omega} \tag{3.11}$$

The heat transfer coefficient α can be calculated by the following equation [49,50]:

$$\alpha = \frac{Nu \cdot \lambda}{d} \tag{3.12}$$

In order to obtain the heat transfer coefficient between the air and water droplets, the following correlation was applied to the spherical water droplets [6,51,52]:

$$Nu = 2 + 0.6Re_d^{1/2} Pr^{1/3} \tag{3.13}$$

The adiabatic expansion efficiency is defined as follows:

$$\eta_{adia} = \frac{W_{adia}}{W_{s\_adia}} \tag{3.14}$$

The efficiency of the near-isothermal expander is defined as follows (which is also called isothermality as the ratio of cycle work output to the ideal isothermal work) [11,53]:

$$\eta_{\text{isoth}} = \frac{W_{\text{n\_iso}}}{W_{\text{iso}}} \tag{3.15}$$

When considering the pump power for spraying water, the near-isothermal expander efficiency is defined as follows:

$$\eta_{\text{isoth2}} = \frac{W_{\text{n\_iso}} - W_{\text{pump}}}{W_{\text{iso}}} \tag{3.16}$$

3.6   Results and discussion

3.6.1   Model validation

To verify that the models developed were working as expected, model validation was carried out. Based on the first stage of the expander, the cylinder diameter was 65 mm, stroke distance 100 mm, top clearance height 70 mm, and rotation speed 500 rpm. The inlet pressure and temperature were 7.03 MPa and 373 K, respectively, while the outlet pressure and temperature were 1.82 MPa and 302.6 K, respectively. The validation results showed that the measured and simulated air mass flow rates were 109.25 and 105.65 kg/h, respectively. Both measured and simulated pressure distributions inside the cylinder are shown in Figure 3.10. TDC point was assumed to be 0 degrees of crank angle and the BDC point was assumed to be 180 degrees, which was the starting point of the cycle simulation of crank angle. Figure 3.10 shows that the simulated results agree well with experimental results. The adiabatic model was validated and it can be further used to analyse the same type of expander with various parameters. As presented above, the

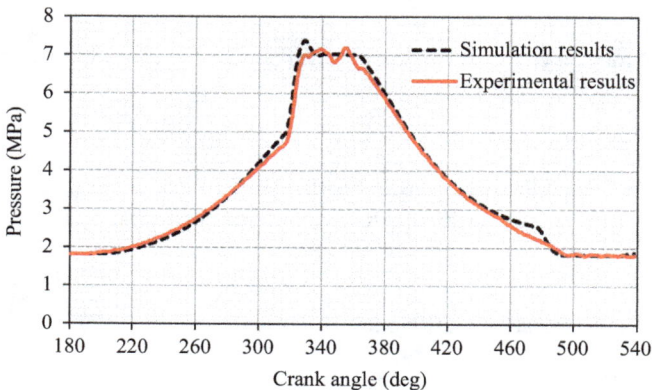

Figure 3.10   Comparison of simulated and experimental results

near-isothermal model was developed by adding the heat transfer equations into the developed adiabatic model. The heat transfer model was also developed and utilized by many other researchers. The near-isothermal model was proved to be effective for the isothermal simulation. Both adiabatic and near-isothermal expansion processes were studied, analysed, and compared in the next section.

3.6.2 Quasi-isothermal expansion analysis

One advantage of the isothermal expander is that it can be operated at a high-pressure ratio through a single stage with a small temperature difference. The pressure ratio was assumed to be 10 in the simulation process. Both adiabatic and near-isothermal expanders were designed and simulated to reveal the relative advantages of the near-isothermal expander. The major parameters of the simulated quasi-isothermal and adiabatic expanders are presented in Table 3.1. Both expanders have the same major parameters including same cylinder diameter, same stroke distance and connecting rod length, same rotation speed, same inlet/outlet pressures, same air inlet temperature, and same air mass flow rate. The main difference of the cylinder structure between the two designs was the top clearance height.

By combining (3.9), (3.10), and (3.11), and choosing average values of the air phase during the near-isothermal expansion process, the Reynolds number was calculated as 90.9, and drag coefficient C_D was 1.2. The Reynolds number less than 200 was within the range of the Ranz–Marshal Correlation (as the Eq. (17) in reference [51]). Furthermore, the Prandtl number and Biot number were calculated, and the Biot number was smaller than 0.1. Thus, the temperature change over the surface of the droplets can be assumed to be uniform.

According to the results shown in Table 3.2, the air mass flow rates were almost the same for both the adiabatic expansion process and the isothermal

Table 3.1   Main parameters of the adiabatic and quasi-isothermal expanders

	Adiabatic expansion	Isothermal expansion
Cylinder diameter (mm)	200	200
Inlet pressure (MPa)	1	1
Air inlet temperature (K)	393.15	393.15
Water temperature (K)	393.15	393.15
Outlet pressure (MPa)	0.1	0.1
Valve lift (mm)	5	4
Inlet period (degree)	46.4	41.4
Rotation speed (rpm)	600	600
Top clearance height (mm)	32	20.5
Outlet period (degree)	74.8	59.1
Stroke distance (mm)	100	100
Connecting rod length (mm)	200	200
Atmosphere temperature (K)	298.15	298.15
Atmosphere pressure (MPa)	0.1	0.1

Table 3.2   Comparison of adiabatic and quasi-isothermal expansion processes

	Adiabatic expansion	Isothermal expansion
Outlet temperature (K)	215.6	375.4
Air mass flow rate (kg/h)	100.38	100.35
Spraying water mass flow (kg/h)	0	100*3
Spraying water droplet diameter (m)	–	30×10^{-6}
Indicated power (kW)	4.55	5.26
Specific work (kJ/kg)	163.18	188.86
Efficiency	85.7%	72.5%/68.5%

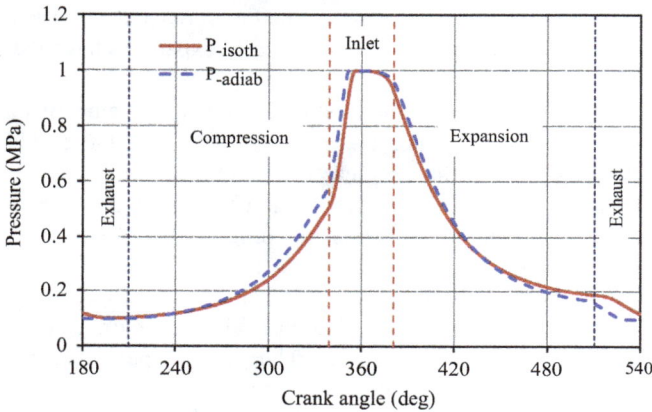

Figure 3.11   Simulated internal pressure vs. crank angle

expansion process. For the isothermal expansion process, the water mass flow rate was estimated to be three times that of the air mass flow rate. Swirl nozzles or pin nozzles can be utilized to spray water droplets [20,54]. The sprayed water droplet diameter from these nozzles is 30 μm. Compared with the adiabatic expander, the power output of the quasi-isothermal expander was increased by 15.6%, and the specific work output increased approximately 15.7%. The isentropic efficiency of the adiabatic expander is 85.7%, in which the mechanical loss is not considered. The efficiency of the near-isothermal expander is 72.5% or 68.5% without or with consideration of the power consumption of the water pump.

The internal pressure distributions of both quasi-isothermal and adiabatic processes along with the crank angle are shown in Figure 3.11, while the stroke distributions (exhaust, compression, inlet, expansion) in one cycle are also labelled in these figures. Both curves were found to be similar. According to (3.8), although the temperature was different between the two processes, the cylinder volume was also different. During the compression process, the pressure of the adiabatic process (P_{adiab}) was a bit higher, and the pressure of the quasi-isothermal process (P_{isoth}) was a bit higher at the later expansion section. This indicates that under the

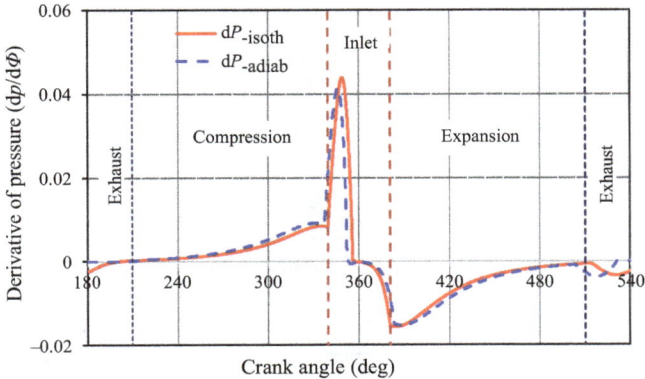

Figure 3.12   Simulated derivative of pressure with respect to crank angle vs. crank angle

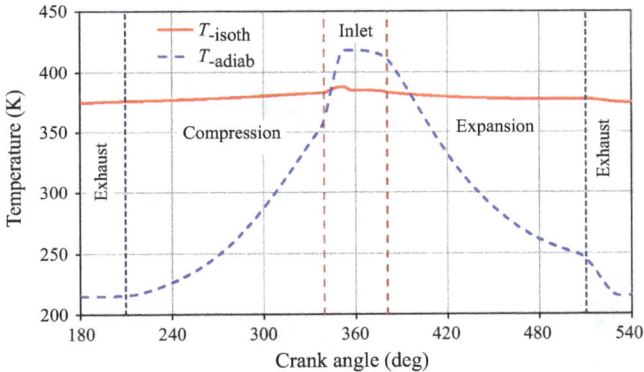

Figure 3.13   Simulated internal temperature vs. crank angle

quasi-isothermal expansion process, the compression section consumes less power and the expansion section generates more power. Thus, the power output is increased. Figure 3.12 shows the derivative of pressure with respect to crank angle as a function of crank angle. It indicates that the most pressure fluctuation occurs during the air inlet period. The air pressure increases quickly as the inlet valve starts to open and decreases quickly after the TDC while the inlet valve starts to close. It also changes much in the expansion process.

Figure 3.13 shows the temperature distribution along with crank angle of both quasi-isothermal and adiabatic processes. It shows that the inlet/outlet temperature difference of one expander cycle is 17.7 K, which is much smaller than the adiabatic process's temperature difference of 177.5 K. It can be calculated that the temperature of quasi-isothermal process is only about 10% of the adiabatic process. Figure 3.14 shows the derivative of the temperature with respect to crank angle of

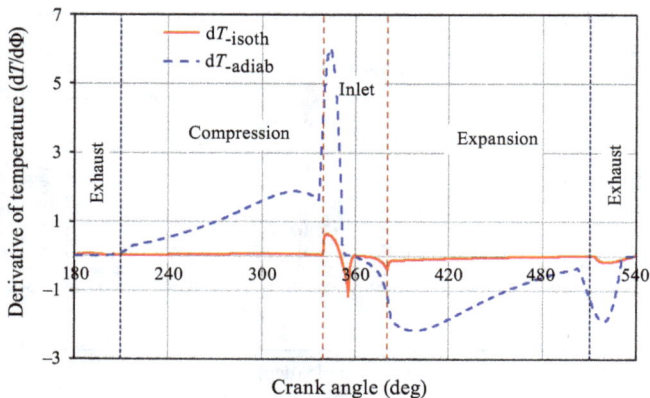

Figure 3.14   *Simulated derivative of temperature with respect to crank angle vs. crank angle*

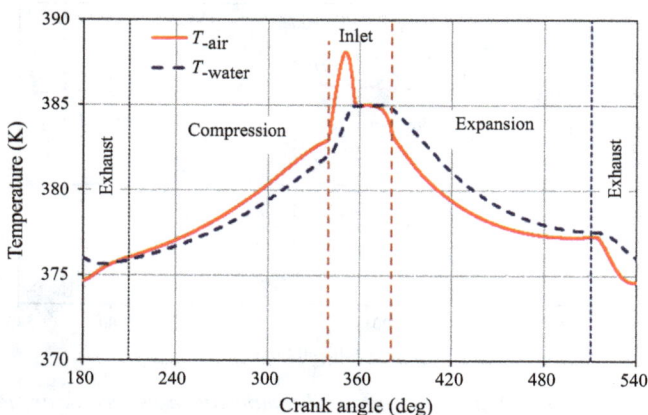

Figure 3.15   *Simulated air and water temperature vs. crank angle*

one cycle. It indicates that the rate of temperature increase is almost zero except during the air intake period. However, during the adiabatic process, the rate of temperature increase changes significantly. For the quasi-isothermal process, the heat capacity of water is much greater than that of the air. Hence, there is relatively sufficient thermal energy from the water to keep the air temperature almost constant. During the air intake period, the high-pressure air runs into the cylinder through the intake valve with a very high speed due to a large pressure difference. Thus, the total air temperature increases, and then decreases quickly as the valve starts to close and the air in the cylinder expands.

Figure 3.15 shows the temperature distribution of both air and water droplets inside the cylinder in one cycle. During the compression section, air temperature

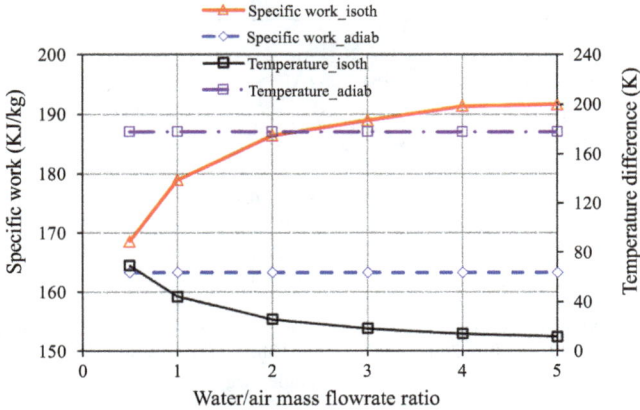

Figure 3.16 Simulated specific work and inlet/outlet temperature difference vs. water/air ratio

was generally higher than that of the water. During the expansion period, the water droplets' temperature was generally higher than the air temperature. The volume ratio (maximum cylinder volume to minimum cylinder volume) of this cylinder was about 5.88, which led to a large temperature increase in the compression period without water spraying. The water droplets helped to slow down this increase. After air intake, air expanded and its temperature decreased simultaneously; water droplets helped to slow down this decrease.

Figure 3.16 reveals the specific work and inlet/outlet temperature difference under various water/air mass flow rate ratio (in the range of 0.5–5) [6]. More specific work was generated through quasi-isothermal expansion. Specific work production increased as the ratio increased, and eventually plateaued. The inlet/outlet temperature difference of quasi-isothermal expansion was much smaller than that of the adiabatic expansion, and the difference tends to plateau as the water/air mass flow rate ratio increases. Figure 3.17 shows the temperature distribution along with one cycle with various water/air mass flow rate ratio. All curves had a similar trend, while the temperature differences between inlet and outlet changed significantly. The difference decreased when the water/air mass flow rate ratio (w/a) increased. This means that with more water sprayed into the cylinder, expansion more closely resembles an isothermal process.

Figure 3.18 shows the specific work production and inlet/outlet temperature difference as a function of the inlet temperature of both air and water, and the water/air mass flow rate ratio of 3. This shows that when the inlet temperature increased, both specific work production and temperature difference increased. The temperature distribution of one cycle under various inlet temperatures is shown in Figure 3.19. All curves have a similar trend, and the inlet/outlet temperature differences are fairly small. This is corresponding to the inlet/outlet temperature difference in Figure 3.18.

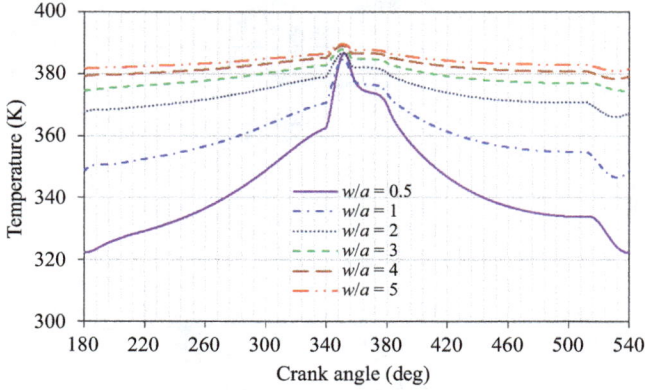

Figure 3.17   Simulated temperature distribution vs. water/air ratio

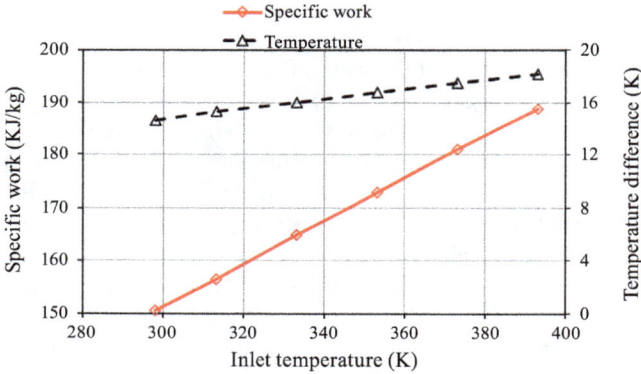

Figure 3.18   Simulated specific work and inlet/outlet temperature difference vs. inlet temperature

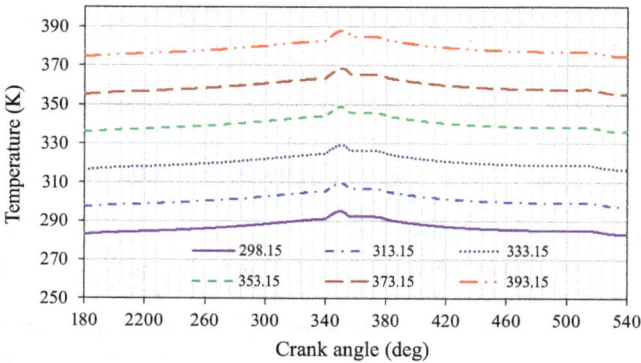

Figure 3.19   Simulated temperature distribution vs. inlet temperature

3.7   Conclusion and remarks

The chapter introduced recent work of I-CAES including system configurations, isothermal compression, and expansion. Various methods have been proposed to achieve isothermal/near-isothermal expansion. I-CAES can achieve relatively high efficiency and high energy density in terms of its high efficiencies of isothermal compressor and expander, and high operation pressure.

A novel reciprocating expander was designed for I-CAES. The operating parameters associated with the expander were measured. Both adiabatic and quasi-isothermal expansion processes were simulated based on the model developed. The adiabatic model was validated using experimental results. A quasi-isothermal model was constructed based on the adiabatic modelling by spraying water droplets into the cylinder. These results indicate that when the expander operates under quasi-isothermal mode, the cylinder height is reduced by 8.7%, the specific work production is increased by 15.7%, and the inlet/outlet temperature difference is only about 10% of that in the adiabatic expansion process.

Acknowledgement

This work was supported in part by the National Science Fund for Distinguished Young Scholars under Grant 51925604 and National Natural Science Foundation of China under Grant 51676181.

References

[1]   Braff WA, Mueller JM, and Trancik JE. Value of storage technologies for wind and solar energy. *Nature Climate Change*. 2016;6:964–9.

[2]   Dodds PE and Garvey SD. Chapter 1: The role of energy storage in low-carbon energy systems. In: Letcher TM, editor. *Storing Energy*. Oxford: Elsevier; 2016. pp. 3–22.

[3]   Arbabzadeh M, Sioshansi R, Johnson JX, and Keoleian GA. The role of energy storage in deep decarbonization of electricity production. *Nature Communications*. 2019;10:3413.

[4]   Mouli-Castillo J, Wilkinson M, Mignard D, McDermott C, Haszeldine RS, and Shipton ZK. Inter-seasonal compressed-air energy storage using saline aquifers. *Nature Energy*. 2019;4:131–9.

[5]   Budt M, Wolf D, Span R, and Yan J. A review on compressed air energy storage: Basic principles, past milestones and recent developments. *Applied Energy*. 2016;170:250–68.

[6]   Qin C and Loth E. Liquid piston compression efficiency with droplet heat transfer. *Applied Energy*. 2014;114:539–50.

[7]   Olabi AG, Wilberforce T, Ramadan M, Abdelkareem MA, and Alami AH. Compressed air energy storage systems: Components and operating parameters – A review. *Journal of Energy Storage*. 2021;34:102000.

[8]   Igobo ON and Davies PA. Review of low-temperature vapour power cycle engines with quasi-isothermal expansion. *Energy*. 2014;70:22–34.

[9]   Odukomaiya A, Abu-Heiba A, Gluesenkamp KR, *et al.* Thermal analysis of near-isothermal compressed gas energy storage system. *Applied Energy*. 2016;179:948–60.

[10]   Arabkoohsar A. Chapter Three: Compressed air energy storage system. In: Arabkoohsar A, editor. *Mechanical Energy Storage Technologies*. London: Academic Press; 2021. pp. 45–71.

[11]   Knowlen C, Williams J, Mattick AT, Deparis H, and Hertzberg A. Quasi-isothermal expansion engines for liquid nitrogen automotive propulsion. SAE International; 1997.

[12]   Heidari M, Mortazavi M, and Rufer A. Design, modeling and experimental validation of a novel finned reciprocating compressor for isothermal compressed air energy storage applications. *Energy*. 2017;140:1252–66.

[13]   Knowlen C, Mattick AT, Bruckner AP, and Hertzberg A. High efficiency energy conversion systems for liquid nitrogen automobiles. SAE International; 1998.

[14]   Li M. Experimental research of internal water-spray cooling in reciprocating compressor. *Fluid Engineering*. 1993;21:5.

[15]   Yan B, Wieberdink J, Shirazi F, Li PY, Simon TW, and Van de Ven JD. Experimental study of heat transfer enhancement in a liquid piston compressor/expander using porous media inserts. *Applied Energy*. 2015;154:40–50.

[16]   Patil VC and Ro PI. Experimental study of heat transfer enhancement in liquid piston compressor using aqueous foam. *Applied Thermal Engineering*. 2020;164:114441.

[17]   Odukomaiya A, Kokou E, Hussein Z, Abu-Heiba A, Graham S, and Momen AM. Near-isothermal-isobaric compressed gas energy storage. *Journal of Energy Storage*. 2017;12:276–87.

[18]   Chen H, Peng Y-h, Wang Y-l, and Zhang J. Thermodynamic analysis of an open type isothermal compressed air energy storage system based on hydraulic pump/turbine and spray cooling. *Energy Conversion and Management*. 2020;204:112293.

[19]   Li PY and Saadat M. An approach to reduce the flow requirement for a liquid piston near-isothermal air compressor/expander in a compressed air energy storage system. *IET Renewable Power Generation*: Institution of Engineering and Technology; 2016. p. 1506–14.

[20]   Qin C, Loth E, Li P, Simon T, and Van de Ven J. Spray-cooling concept for wind-based compressed air energy storage. *Journal of Renewable and Sustainable Energy*. 2014;6:043125.

[21]   Qin C, Saunders G, and Loth E. Offshore wind energy storage concept for cost-of-rated-power savings. *Applied Energy*. 2017;201:148–57.

[22]   Bollinger B. Demonstration of isothermal compressed air energy storage to support renewable energy production. SustainX, Inc.; 2015. p. 50.

[23]   Zhang X, Chen H, Xu Y, *et al.* Distributed generation with energy storage systems: A case study. *Applied Energy*. 2017.

[24] Zhang C, Shirazi FA, Yan B, Simon TW, Li PY, and Van de Ven J. Design of an interrupted-plate heat exchanger used in a liquid-piston compression chamber for compressed air energy storage. *ASME 2013 Heat Transfer Summer Conference collocated with the ASME 2013 7th International Conference on Energy Sustainability and the ASME 2013 11th International Conference on Fuel Cell Science*, Engineering and Technology, 2013.

[25] Zhang C, Yan B, Wieberdink J, *et al.* Thermal analysis of a compressor for application to compressed air energy storage. *Applied Thermal Engineering*. 2014;73:1402–11.

[26] Guanwei J, Weiqing X, Maolin C, and Yan S. Micron-sized water spray-cooled quasi-isothermal compression for compressed air energy storage. *Experimental Thermal and Fluid Science*. 2018;96:470–81.

[27] Heo JY, Kim MS, Baik S, Bae SJ, and Lee JI. Thermodynamic study of supercritical CO2 Brayton cycle using an isothermal compressor. *Applied Energy*. 2017;206:1118–30.

[28] Mohammadi-Amin M, Jahangiri AR, and Bustanchy M. Thermodynamic modeling, CFD analysis and parametric study of a near-isothermal reciprocating compressor. *Thermal Science and Engineering Progress*. 2020;19: 100624.

[29] Busarov SS, Tretyakov AV, Sherban KV, and Balakin PD. Thermal stress state of the parts of quasi-isothermal long-stroke low flow stages in reciprocating compressors. *Procedia Engineering*. 2016;152:303–8.

[30] Van de Ven JD and Li PY. Liquid piston gas compression. *Applied Energy*. 2009;86:2183–91.

[31] Cicconardi SP, Jannelli E, and Spazzafumo G. A thermodynamic cycle with a quasi-isothermal expansion. *International Journal of Hydrogen Energy*. 1998;23:209–11.

[32] Kim YM and Favrat D. Energy and exergy analysis of a micro-compressed air energy storage and air cycle heating and cooling system. *Energy*. 2010;35:213–20.

[33] Woodland BJ, Braun JE, Groll EA, and Horton WT. Performance benefits for organic Rankine cycles with flooded expansion. International Refrigeration and Air Conditioning Conference, Purdue, USA: Publications of the Ray W. Herrick Laboratories; 2010.

[34] Park J, Ro PI, He X, and Mazzoleni AP. Analysis, fabrication, and testing of a liquid piston compressor prototype for an ocean compressed air energy storage (OCAES) system. *Marine Technology Society Journal*. 2014;48:86–97.

[35] Moradi J, Shahinzadeh H, Khandan A, and Moazzami M. A profitability investigation into the collaborative operation of wind and underwater compressed air energy storage units in the spot market. *Energy*. 2017;141:1779–94.

[36] Bell IH, Lemort V, Groll EA, Braun JE, King GB, and Horton WT. Liquid-flooded compression and expansion in scroll machines – Part I: Model development. *International Journal of Refrigeration*. 2012;35:1878–89.

[37] James NA, Braun JE, Groll EA, and Horton WT. Semi-empirical modeling and analysis of oil flooded R410A scroll compressors with liquid injection for

use in vapor compression systems. *International Journal of Refrigeration.* 2016;66:50–63.

[38]   Xu Z, Lu Y, Wang B, Zhao L, and Xiao Y. Experimental study on the off-design performances of a micro humid air turbine cycle: Thermodynamics, emissions and heat exchange. *Energy.* 2021;219:119660.

[39]   SIEMENS. Fast wet compression for Siemens gas turbines. SIEMENS; 2015.

[40]   SIEMENS. Wet compression. 2020.

[41]   Zhang X, Xu Y, Xu J, Xue H, and Chen H. Study of a single-valve reciprocating expander. *Journal of the Energy Institute.* 2016;89:400–13.

[42]   He W and Wang J. Optimal selection of air expansion machine in compressed air energy storage: A review. *Renewable and Sustainable Energy Reviews.* 2018;87:77–95.

[43]   Marvania D and Subudhi S. A comprehensive review on compressed air powered engine. *Renewable and Sustainable Energy Reviews.* 2017;70:1119-30.

[44]   Dellicompagni P, Saravia L, Altamirano M, and Franco J. Simulation and testing of a solar reciprocating steam engine. *Energy.* 2018.

[45]   Rahbar K, Mahmoud S, Al-Dadah RK, Moazami N, and Mirhadizadeh SA. Review of organic Rankine cycle for small-scale applications. *Energy Conversion and Management.* 2017;134:135–55.

[46]   Wasbari F, Bakar RA, Gan LM, Tahir MM, and Yusof AA. A review of compressed-air hybrid technology in vehicle system. *Renewable and Sustainable Energy Reviews.* 2017;67:935–53.

[47]   Béguin C, Étienne S, and Pettigrew MJ. Effect of dispersed phase fraction on the drag coefficient of a droplet or a bubble in an idealized two-phase flow. *European Journal of Mechanics – B/Fluids.* 2017;65:339–49.

[48]   Kelbaliyev G and Ceylan K. Development of new empirical equations for estimation of drag coefficient, shape deformation, and rising velocity of gas bubbles or liquid drops. *Chemical Engineering Communications.* 2007;194:1623–37.

[49]   Lin S and Zhao G. Thermodynamical research of reciprocating compressor spraying water inside for colling. *Journal of Engineering Thermophysics.* 1987;8:3.

[50]   Sirignano WA. *Fluid Dynamics and Transport of Droplets and Sprays.* New York: Cambridge University Press; 2010.

[51]   Ranz WE and Marshall WR. Evaporation from drops: Part 1. *Chemical Engineering Progress.* 1952;48:141–6.

[52]   Yuen MC and Chen LW. Heat-transfer measurements of evaporating liquid droplets. *International Journal of Heat and Mass Transfer.* 1978;21:537–42.

[53]   Chen H, Ding Y, Li Y, Zhang X, and Tan C. Air fuelled zero emission road transportation: A comparative study. *Applied Energy.* 2011;88:337–42.

[54]   Chaker MA. Key parameters for the performance of impaction-pin nozzles used in inlet fogging of gas turbine engines. *Journal of Engineering for Gas Turbines and Power.* 2006;129:473–7.

Chapter 4

Improving the efficiency of A-CAES systems by preconditioning discharge air stream

Mehdi Ebrahimi[1], David Brown[2], David S-K. Ting[1], Rupp Carriveau[1] and Andrew McGillis[2]

This chapter explores the effects of variations in cavern air quality on the overall performance of adiabatic compressed air energy storage (A-CAES) systems. Components of A-CAES systems interact in dynamic ways and often have compounding effects. A small change in the state of one of the key process parameters can therefore propagate through and alter the status of the entire system. In case of a discharging A-CAES system, the key process parameters (mass flow rate, temperature, pressure, and humidity) of the inlet air stream will therefore significantly influence the performance of an A-CAES plant. However, these parameters—particularly temperature and relative humidity—are not generally controlled for an A-CAES facility. They are instead allowed to settle to natural values which depend on operational frequency, ambient conditions, and cavern design choices. It would however be possible to control these parameters, at the cost of the infrastructure (heaters, coolers, humidifiers, regulators, etc.). This chapter analyzes the effect of variations in these parameters on the performance of A-CAES systems by thermodynamic simulation of a 100 MW plant with three stages of expansion. For a fair comparison in between generated models, the atmospheric condition, the volume of the hot thermal store, and the efficiency of expanders were fixed for all cases. Simulations were performed for systems with cavern pressures between 4 and 8 MPa, where the ratios between hot water mass flow and air mass flow were between 0.65 and 0.85. The obtained result shows that the maximum reachable power and efficiency in the considered ranges are 117 MW and 80.3%, respectively. Also, the maximum and the minimum accessible energies are 482 and 324 MWh, respectively, with corresponding discharge durations of 4.5 and 3.4 h. The optimum performance of the system was observed when the ratio of hot water to air mass flow was 0.8. Moreover, this work revealed that every 25% increase in

[1]Turbulence and Energy Lab., Ed Lumley Centre for Engineering Innovation, University of Windsor, Windsor, Ontario, Canada
[2]Hydrostor Inc., Toronto, Canada

humidity percentage can increase the total expander power by about 0.1%. The output of this work can also be used for the management of the thermal systems according to the cavern air property.

Keywords: Compressed air energy storage; A-CASE performance; cavern air quality

Nomenclature

C_p	Specific heat at constant pressure
e	Specific exergy (J/kg)
E	Energy (J)
E_D	Exergy destruction (J)
E_x	Exergy (J)
h	Enthalpy (J)
m	Mass flow rate (kg/s)
P	Pressure (kPa)
Q	Heat (J)
R	Gas constant (J/(kg K))
s	Entropy (J/K)
T	Temperature (K)
W	Work (J)
W_{exp}	Expander's power (kW)

Greek letters

γ	Polytropic exponent
η	Pressure loss constant

Subscripts and superscripts

cold	Cold medium
f	Final state
hot	Hot medium
i	Initial state
in	Inlet

Abbreviations

A-CAES	Adiabatic compressed air energy storage
CAES	Compressed air energy storage

EX	Turbo expander
GN	Generator
HPD	High pressure line
HTK	Hot tank
HX	Heat exchanger
IPD	Intermediate pressure line
LPD	Low pressure line
RMFR	Relative mass flow rate
THT	From the hot tank

4.1 Introduction

In the fight against climate change and pollution, it is necessary to increase energy production without negatively impacting the environment. This has led to more renewable energy generation and the development of new renewable technologies [1,2]. However, the adoption of renewable energy has been slowed by the uncertainty inherent in the intermittent nature and weather dependency of almost all renewable energy sources. This uncertainty can be removed through the deployment of energy storage assets; however, storage technology has thus far lagged generation. Compressed air energy storage (CAES) technology has recently gained a lot of worldwide attention as a potential solution to this issue due to its high reliability and low environmental impact [3]. In CAES technology, surplus electricity is converted to high-pressure compressed air, and is released and reconverted when electricity is needed. This technology has the potential to offer a very low cost per kWh at large scale and can also be flexibly sited. There is therefore a potential to co-locate with large-scale renewable energy systems to provide peak shaving.

The fundamental concept of CAES technology has been in use for more than four decades. In 1978, the first commercial-scale cavern-based CAES was constructed in Huntorf, Germany, with the capacity of 321 MW. This plant can store pressurized air in two caverns with a total volume of 310,000 m^3. Normally, the compression phase takes about 12 h and consumes about 750 MWh of energy. The discharge cycle can then operate for more than 3 h at the rated output power. However, in traditional CAES plants such as this one, air is injected into natural gas burners to heat it up before entering the expanders, thus resulting in CO_2 emissions. In 1991, a second large CAES plant was built with a capacity of 110 MW in McIntosh, USA. In this plant, the cavern volume is 19 million cubic meters, and the nominal discharge cycle can be up to 26 h. In this project, a portion of waste heat in the compression phase is recovered, resulting in a fuel consumption rate about 25% less than that of the Huntorf plant [4].

Currently, there are a couple of other CAES plants under operation, construction, or in the study phase. Generally, these new plants aim to improve upon the technology employed at the Huntorf or McIntosh facilities. For example, the world's first research underwater CAES (UWCAES) plant was constructed in Toronto, Canada, by Hydrostor in 2015 with a capacity of 0.7 MW [5]. A feasibility study for a 160 MW cavern-based CAES plant near Saskatchewan and Alberta border in Canada was accomplished in 2018 [6]. The aim of this plant is to combine CAES technology with power grids for peak shaving. A CAES demonstration plant is in the development stage, at Saxony-Anhalt in Germany. The capacity of this plant will be 360 MWh with an output power of 90 MW, aiming for a cycle efficiency of about 70% [7].

Almost all of the CAES plants currently in operation do not make use of the heat of compression, instead exhausting this heat to atmosphere. This then requires the addition of heat in the discharge cycle, which is usually provided by burning fossil fuels. To deal with this environmental issue, some innovations on traditional CAES technology have been recently introduced.

One of these new concepts is adiabatic CAES (A-CAES). Adiabatic here meaning that the goal is to recover the heat of compression and reuse it during expansion operations, thus adding no external heat to the process and requiring no fuel. In A-CAES, the heat of compression is stored and is reused to heat up the compressed air during the discharging cycles [8,9]. Currently, there is only one A-CAES plant under operation in the world in a salt mine in Goderich, Canada [10].

In terms of theoretical studies, Liu *et al.* [11] simulated a combined cycle CAES system and thermodynamically analyzed the performance of the system. They reported that the overall efficiency was improved by about 10% comparing to the original CAES technology. They showed that the heat obtained from the compressor intercoolers can be utilized for improving system performance by keeping the steam part of the plant on hot standby. Li *et al.* [12] introduced a CAES-based trigeneration system which allows trigeneration of electrical, cooling, and heating power during the discharge cycle. Tada and Yoshida [13] simulated airflow of the Kami-Sunagawa CAES plant in Japan and calculated the spatial distribution of air pressure and temperature in the cavern. Raju and Khaitan [14] modeled the thermodynamic behavior of air in the cavern of a CAES plant by applying mass and energy conservation equations. They proposed a flexible method for the calculation of the heat transfer coefficient. They calibrated the developed model with the real data from the Huntorf CAES plant. Sciacovelli *et al.* [15] proposed an A-CAES system with adding a thermal energy storage system to the original CAES technology. In their model, no external heat was added to the system. By this modification, the round-trip efficiency improved to about 69%. Zhang *et al.* [16] used parallel air storage tanks to study the effect of adiabatic air storage chamber on the thermodynamic performance of CAES systems. This showed that different air storage models can result in different characteristics of compression and expansion processes. Najjar and Jubeh [17] showed that the performance of a CAES can be increased by adding a humidification system to the process. The saturator in the process is used for the increase of the moisture content of air before entering combustor. They assumed the liquid water in saturator is warmed by the compression heat.

Despite much effort by both public and private entities, the early-stage development of this technology has been slow with only a few large-scale plants realized over the past few decades. As discussed, there is only one operating A-CAES plant in the world—the one in Goderich, Canada. By studying the influence of variations in the main discharge process parameters in this technology on the performance of the system as a whole, this work aims to enhance our knowledge about CAES technology and specifically A-CAES systems. From previous studies [18,19], we know that in CAES systems, the interaction of equipment on each other is not linear and a small change in the status of the one of the system components can significantly alter the efficiency of the entire system. The change in the system status can also alter the performance of the thermal subsystems in A-CAES technology. To improve our knowledge about the thermal subsystem behavior, the thermal performance and the thermal fluid mass flow rate at various system statuses are also studied. The outcome of this work can be used as a guideline for the design of thermal system in A-CAES systems.

4.2 A-CAES technology description

A-CAES technology can be considered as a variation of the gas turbine cycle where the compression and expansion cycles operate independently. The focus of this study is the discharge cycle of an A-CAES facility, which has been relatively understudied. Studies about the charge cycle can be found in other works such as [20–22]. The schematic flow diagram of an A-CAES system is depicted in Figure 4.1. Generally, in A-CAES systems, air is pulled into the system via compressors and is transferred to the air storage which is usually an underground

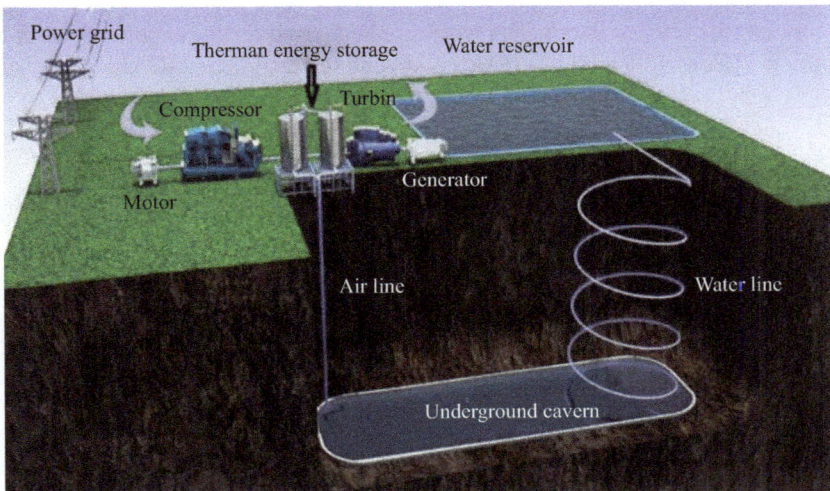

Figure 4.1   Process flow diagram for A-CAES concept

cavern. In this process, a plenty of heat is generated. To recover the heat of compression, and to cool down the air prior to its entrance into the next compression
phase, heat exchangers are installed after some or all the compressors which
transfer the heat to a thermal fluid (usually water). The heated thermal fluid is then
stored in a large tank, usually pressurized. In the discharge cycle, the compressed
cold air returns from the cavern to the system. The air is then heated by the stored
hot thermal fluid using the same heat exchangers as on the charge cycle. The hot,
pressurized air is then used to turn turbines and generate electricity. The cold
thermal fluid is stored in a separate, cold tank. It is worth mentioning that system
and machine cooling loads could also be reduced dramatically by cavern air pre-
conditioning on discharge if the plant's waste heat is used to heat the cavern air.

4.3   Methodology

4.3.1   *Thermodynamic simulation of A-CAES systems*

In this work, a thermodynamic analysis is applied to a simulation of an A-CAES
plant with a capacity of 100 MW (Figure 4.2). The thermodynamic cycle method is
used for the simulation of the A-CAES system. It has been shown that this method
provides higher accuracy than others since the thermodynamic specifications of the
model are well understood [23,24]. In the analysis, the following assumptions have
been applied:

- The volume and the temperature of the hot water for all cases are the same.
- The potential and the kinetic energy of the working fluids are negligible.
- The isentropic efficiencies of all expanders are 89%.
- No chemical reactions occur.
- The atmospheric pressure for all cases is 100 kPa.

Bounds for process parameters were selected based on our experience with the
operation of Toronto Island UWCAES plant and the Goderich A-CAES plant, such

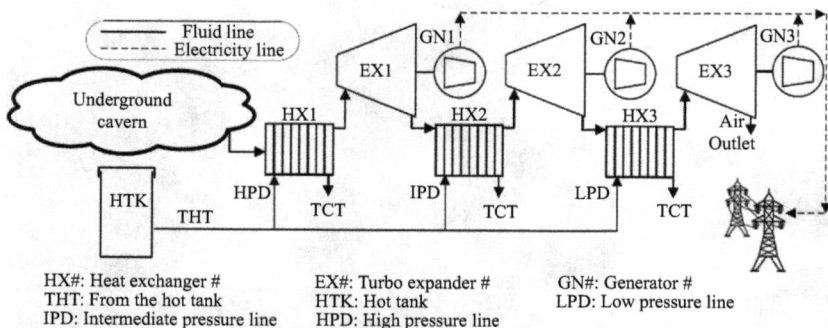

HX#: Heat exchanger #   EX#: Turbo expander #   GN#: Generator #
THT: From the hot tank   HTK: Hot tank   LPD: Low pressure line
IPD: Intermediate pressure line   HPD: High pressure line

*Figure 4.2   The schematic diagram of the applied A-CAES for the thermodynamic
analysis of a 100 MW plant*

Table 4.1   The maximum and minimum values for the understudied parameters

Parameters	Unit	Minimum value	Maximum value
Cavern pressure	kPa	4,000	8,000
Cavern temperature	°C	5	25
Air humidity at the first expander's inlet	%	~0	~100
Hot water to air mass flow rate	–	0.65	0.85

Table 4.2   Air thermodynamic properties for two different states of the system with the air humidity of 60% and 90% at the inlet of the first expander

State point	Fluid	State point 1 with 60% humidity			State point 2 with 90% humidity		
		T (°C)	P (kPa)	Mass (kg/s)	T (°C)	P (kPa)	Mass (kg/s)
Cavern outlet	Air	15.0	6,000	270.10	10.0	7,000	270.10
Wellhead	Air	8.6	5,559	270.10	3.6	6,465	270.10
EX1 inlet	Air	184.9	5,534	270.10	182.6	6,440	270.10
EX1 outlet	Air	50.2	1,384	270.10	48.29	1,610	270.10
EX2 inlet	Air	192.1	1,352	270.10	191.7	1,579	270.10
EX2 outlet	Air	56.7	338	270.10	56.3	395	270.10
EX3 inlet	Air	194.4	298	270.10	194.3	354	270.10
EX3 outlet	Air	83.6	100	270.10	68.6	100	270.10
THT	Water	213.0	800	189.10	213.0	800	189.10
HPD	Water	23.6	798	62.97	18.59	798	62.97
IPD	Water	65.2	799	61.84	63.27	799	61.84
LPD	Water	76.64	798	64.29	76.27	798	64.29
TCT	Water	55.24	798	189.10	7,985	798	189.10

that the system simulated is one that has a useful round-trip efficiency and could be built at reasonable cost. The result was that simulations were performed for systems with cavern pressures between 4 and 8 MPa, cavern air temperatures between 5 °C and 25 °C, and relative air humidities between 0% and 100%, where the ratios between hot water mass flow and air mass flow were between 0.65 and 0.85 (Table 4.1). The status of the system at any point can then be determined using the developed thermodynamic model. In Table 4.2, the thermodynamic state of the air and water streams are presented at the inlets and outlets of every major piece of equipment, and how they respond to different cavern outlet conditions for two sample state points.

4.3.1.1   System modeling

Thermodynamic evaluation of the A-CAES model has been carried out by simulating the different components of the system based on energy and mass balances. The modeling details of the system components are described here.

Humid air model
In this work, air is treated as an ideal gas and the specific heat of air is considered as a function of temperature and the ratio of specific heats. A logarithmic polynomial function [25] was applied for the evaluation of the specific heats of working fluids as follows:

$$C_p(T) = a + b\,T + c\,T^2 + d\,T^3 + e\,T^4 + f\,T^5 \tag{4.1}$$

where T is the temperature in Kelvin and a, b, c, d, e, and f are the polynomial coefficients. The enthalpy of humid air at any temperature then can be calculated as

$$h = C_{p-\text{air}}(T) + \left(2,500 + c_{p-h_2o}(T)\right) \tag{4.2}$$

Heat exchangers
Heat exchangers are integral parts of A-CAES systems as they serve as the connection/interface point for the thermal and air subsystems. A well-designed thermal management system has the potential to improve the performance of A-CAES cycles by up to 80% relative to traditional CAES systems [26]. If heat loss to the environment and transient effects (i.e., heating of metals, residence time of fluids in the exchangers) are ignored, then the rate of heat transfer in a heat exchanger can be described by

$$\dot{Q} = \dot{m}_{\text{hot}}C_{p,\text{hot}}\left(T_{\text{hot,in}} - T_{\text{hot,out}}\right) = \dot{m}_{\text{cold}}C_{p,\text{cold}}\left(T_{\text{cold,out}} - T_{\text{cold,in}}\right) \tag{4.3}$$

where $\dot{m}$ is the mass flow rate (kg/s), T is the temperature (K), and $C_{p-\text{air}}$ is the specific heat capacity at constant pressure (J/(K kg)).

The pressure drop can also be calculated with an empirical equation [27] as

$$P_{\text{loss}} = \frac{0.0083\eta}{1 - \eta}P_{\text{in}} \tag{4.4}$$

where P_{in} is the pressure at the heat exchanger inlet. According to the defined system, in this work it can be between 4 and 10 kPa. It should be noted that in this work water is used as the thermal fluid due to its high thermal capacity and low cost. Generally, water is unstable and difficult to manage at high temperatures. Therefore, in each project according to the thermodynamic condition of heat exchanges, proper thermal fluid should be selected. At the scale and process parameters simulated, water will always be the thermal fluid of choice for economic reasons—use of a different fluid would result in a significantly higher price per kWh stored.

Expanders
To calculate the power generated by turbines, the exit temperatures of turbines need to be calculated. Considering isentropic expansion in the expanders and air behaving as an ideal gas with a constant specific heat, the relation between the pressure and temperature at any stage of expansion can be calculated by

$$T_{\text{f}} = T_{\text{i}}\left(\frac{P_{\text{f}}}{P_{\text{i}}}\right)^{\frac{\gamma-1}{\gamma}} \tag{4.5}$$

In this equation, i and f are the initial and final status of air, respectively, and γ is the specific heat ratio. The expander power at any stage can be determined as

$$\dot{W}_{\text{exp.}} = \frac{\dot{m}RT_i}{\eta_{\text{exp.}}}\left(1 - \left(\frac{P_f}{P_i}\right)^{\frac{\gamma-1}{\gamma}}\right)\frac{\gamma}{\gamma-1} \tag{4.6}$$

For a fair comparison between models, it is assumed that the isentropic efficiency of the turbine is 89% at any stage [28]. Then, the final expansion temperature can be determined by

$$T_{\text{exp}} = T_i\left(1 - \eta\left(1 - \left(\frac{P_f}{P_i}\right)^{\frac{\gamma-1}{\gamma}}\right)\right) \tag{4.7}$$

System efficiency and heat efficiency
The proposed A-CAES system has two different energy inputs: air stream energy and hot water stream energy. It also has two energy outputs: the electrical energy of expanders and the heat energy of water leaving the system. To analyze the discharge cycle and investigate the effect of air property changes on the system performance, two efficiencies including discharge cycle efficiency and heat recovery efficiency are defined in this work. These two efficiencies are defined as follows:

$$\text{Cycle efficiency} = \frac{\dot{E}_{\text{Turb1}} + \dot{E}_{\text{Turb2}} + \dot{E}_{\text{Turb3}}}{\dot{E}_{\text{CavernOutlet}} + \dot{E}_{\text{HotW\_in}}} \tag{4.8}$$

$$\text{Heat efficiency} = \frac{\dot{E}_{\text{HotW}_i\text{n}} - \dot{E}_{\text{HotW}_o\text{ut}}}{\dot{E}_{\text{HotW\_in}}} \tag{4.9}$$

where $\dot{E}_{\text{Turb}}$ is the kth expander's energy, $\dot{E}_{\text{CavernOutlet}}$ air stream energy at the outlet of the cavern, and $\dot{E}_{\text{HotW}_i\text{n}}$ the energy of hot water at the inlet of the heat exchangers.

Thermodynamic analysis
The mass, energy, and exergy balance of the system components can be written as

$$\sum \dot{m}_{\text{in}} - \sum \dot{m}_{\text{out}} = 0 \tag{4.10}$$

$$\dot{Q} - \dot{W} = \sum \dot{m}_{\text{out}}h_{\text{out}} - \sum \dot{m}_{\text{in}}h_{\text{in}} \tag{4.11}$$

$$\dot{Q} - \dot{W} = \sum \dot{m}_{\text{out}}e_{\text{out}} - \sum \dot{m}_{\text{in}}e_{\text{in}} + \dot{E}_D \tag{4.12}$$

where $\dot{W}$ is the rate of work leaving the control volume (kW), e is the specific exergy (kJ/kg), and $\dot{E}_D$ is the exergy destruction rate. Also, exergy is the maximum amount of useful energy when the system is in mechanical and thermal equilibrium with the surrounding environment.

The overall exergy at any state point can be presented by thermomechanical and chemical exergy. Since there is no chemical reaction in the modeled A-CAES,

Table 4.3   The assumption of parameters considering for
calculation of system status

Component	Parameters
EX1	$\eta_{\text{Isentropic}} = 89\%$
HX1	Air pressure loss = up to 14 kPa (see (4.4))
	Water pressure loss $\cong$ 5 kPa
	Water inlet temperature = 213 °
	Hot water mass flow rate portion = 0.333
Ex2	$\eta_{\text{Isentropic}} = 89\%$
HX2	Air pressure loss = up to 11 kPa (see (4.4))
	Water pressure loss $\cong$ 5kPa
	Water inlet temperature = 213 °
	Hot water mass flow rate portion = 0.327
EX3	$\eta_{\text{Isentropic}} = 89\%$
HX3	Air pressure loss = up to 9 kPa (see (4.4))
	Water pressure loss $\cong$ 5kPa
	Water inlet temperature = 213 °
	Hot water mass flow rate portion = 0.340
Well	Air pressure loss $\cong$ 7.6%of the cavern presure

the total exergy of system at any point can be calculated by

$$\dot{Ex} = \dot{m}e \tag{4.13}$$

The specific exergy can be expressed as

$$e = (h - h_o) - T_o(s - s_o) \tag{4.14}$$

where s_o is the specific entropy at the ambient environmental condition. In real facilities, main system parameters (T, P, and mass flow rates) at any point can be obtained from sensors. For calculating specific entropy, enthalpy, and exergy, some more information is required. These additional details for the current work are shown in Table 4.3.

4.4   Results and discussion

This work theoretically analyzed the effect of cavern air temperature and pressure, expander pressure ratios, ratio of air mass flow rate to hot water mass flow rate (RMFR), and relative humidity of the air stream from the cavern on the performance of A-CAES systems by thermodynamic simulation of a wide range of models. The boundaries of the design conditions considered are given in Table 4.1. The applied model has a 100 MW rated power, the size of water hot tank is 3,000 m³, and the water is at a temperature of 213 °C. The outlet pressure of the system is fixed at 100 kPa and the system discharges until it runs out of hot water. In Figure 4.3, air exergy and generated power by expanders are

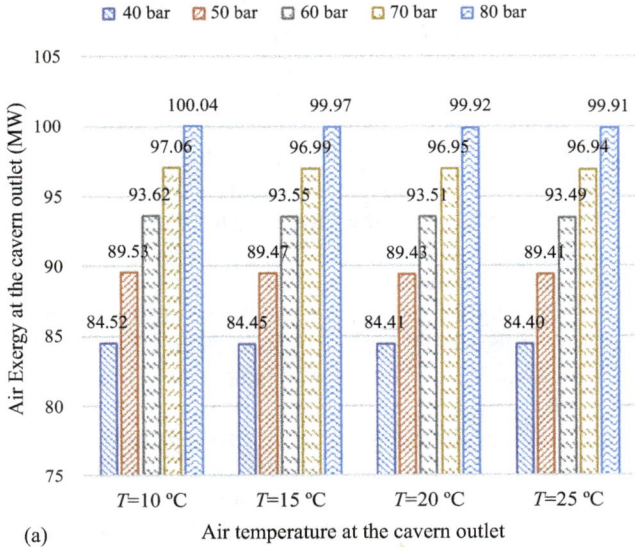

(a)

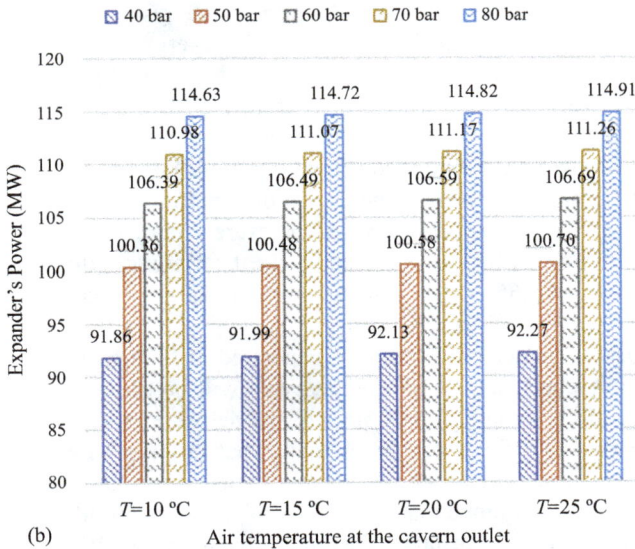

(b)

Figure 4.3   Air exergy and the total generated power by expanders at various cavern pressure and temperature

compared at various cavern temperatures and pressures with the water to air mass flow rate ratio of 0.8. This figure shows that the effect of air pressure change is greater that the temperature change. It also can be observed that at lower pressure, the effect of pressure change is more significant than higher pressures. For example, when at 10 °C air pressure changes from 40 to 50 bar, turbine's total power increases by 9.2%, while when pressure changes from 70 to 80 bar, the output power change is 3.3%. This makes sense as the energy extracted by a turbine depends on the ratio of inlet/outlet conditions rather than their absolute magnitudes.

In Figure 4.4, the efficiency of the discharge cycle at various operating conditions is shown. This figure shows that by increasing cavern air pressure, the efficiency is improved; however, there are diminishing returns. It also shows that with every 5 °C increase in cavern air temperature, the efficiency is improved by about 0.1%.

In Figure 4.5, air exergy at the outlet of the cavern is shown at different system pressures. The corresponding expander power values for different hot water to air mass flow rates at various operating conditions are shown in Figure 4.6. From this figure, it can be concluded that at higher air mass flow rate, greater power is accessible. However, it should be considered that at higher air mass flow rate, cavern pressure reaches to its minimum level faster and lower discharge time will be available.

In Figure 4.7, hot water exergy at the inlet of the system for a mass flow rate of 230 kg/s is shown. The corresponding outlet exergy of the water stream at various RMFRs are shown in Figure 4.8. According to this figure, more energy is transferred between air and water lines at low RMFRs. The heat efficiency (Figure 4.9) and the discharge cycle efficiency (Figure 4.10) can then be determined based on the values obtained. From Figure 4.9, it can be observed that at lower RMFRs, higher heat efficiencies are obtainable. It can also be seen that heat efficiency improves as cavern pressure increases. However, from Figure 4.10, it can be observed that unlike heat efficiency, the total cycle efficiency at RMFR = 0.65 is the lowest value and the maximum value occurs at RMFR = 0.8.

As discussed, higher power can be generated at higher air mass flow rates. However, at this condition, the discharge cycle will be shorter. To have a better understanding of the situation in Table 4.4, the delivered energy for each case is presented. From this table, it can be observed that in terms of energy, cycles are more efficient at higher pressures and lower air mass flow rates. It should be noted that in this table, it is assumed that expander's efficiencies for all cases are 89%. In practice, maintaining a constant efficiency for lower RMFRs requires more expensive expanders. Therefore, a cost-benefit analysis for determining the right expander for each project is essential.

The variation in air humidity is another factor that can affect system performance. In Figure 4.11, the relationship between relative and absolute humidity of the air stream at various temperatures is shown. Water molar fractions from zero to 3×10^{-4} were investigated in order to capture the full range of

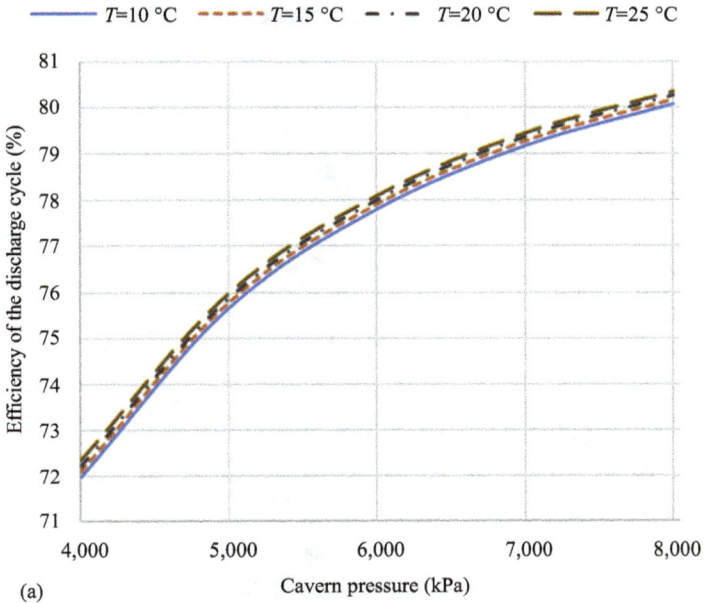

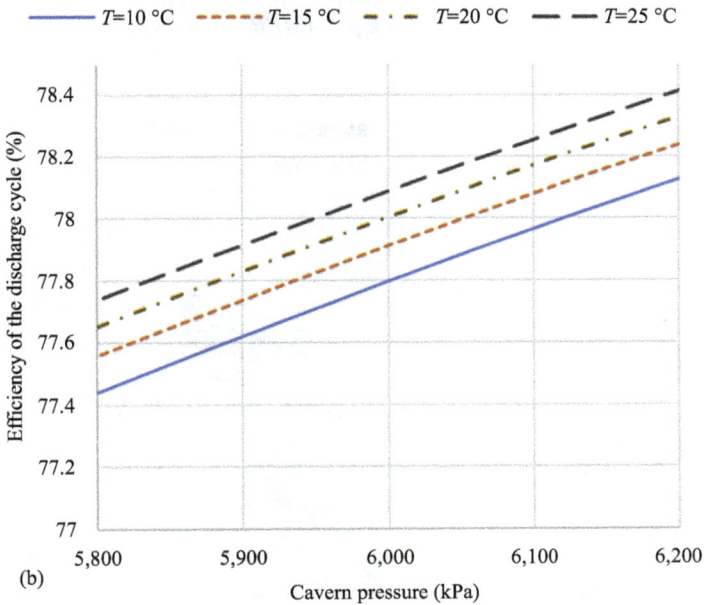

Figure 4.4   Efficiency of the discharge cycle at various cavern air pressure and temperature

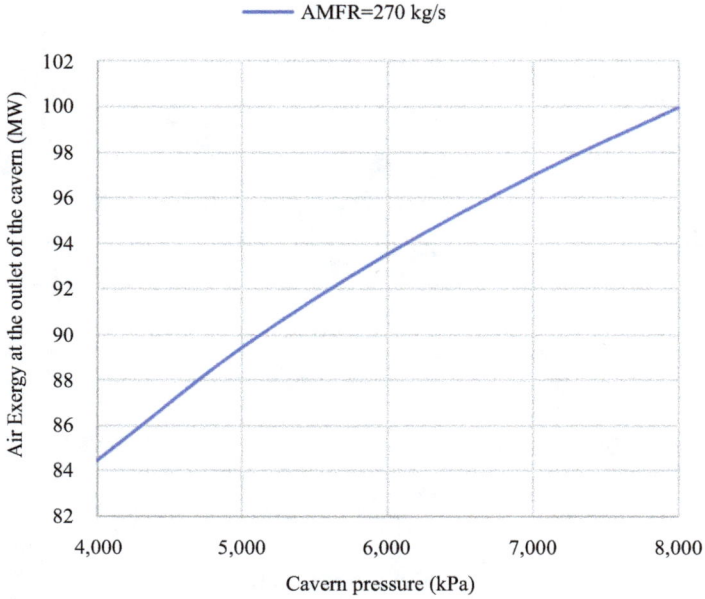

Figure 4.5   *Air exergy at the outlet of the cavern for various cavern pressure at the air mass flow rate of 270 kg/s. AMFR: air mass flow rate*

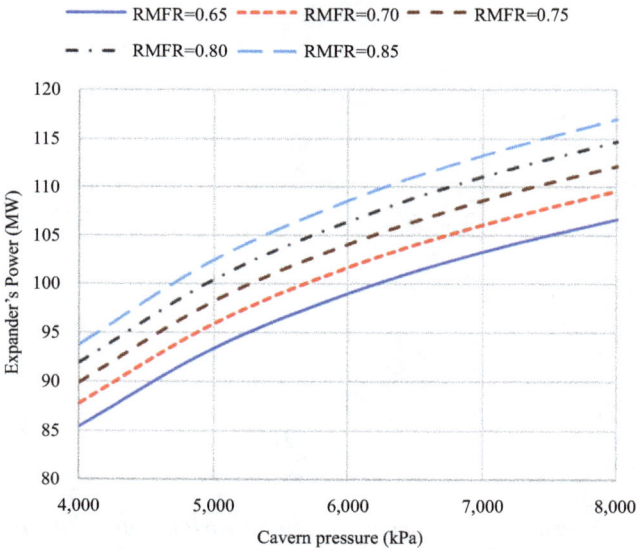

Figure 4.6   *Total generated power by expanders at various hot water to air mass flow rate*

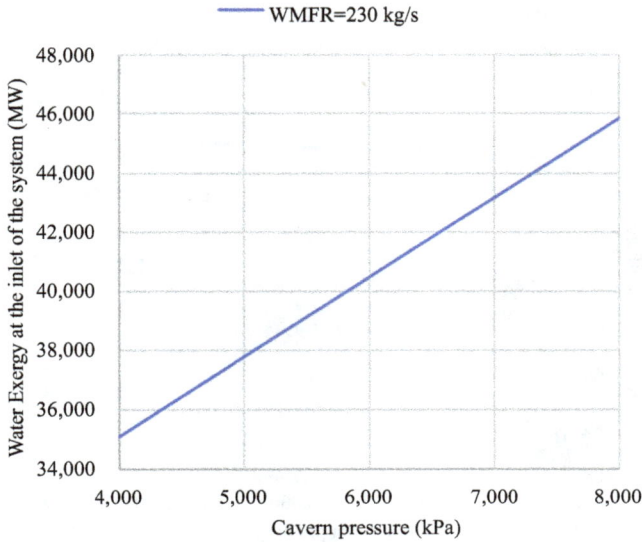

Figure 4.7   Air exergy at the outlet of the cavern for various cavern pressure at the air mass flow rate of 270 kg/s. WMFR: water mass flow rate

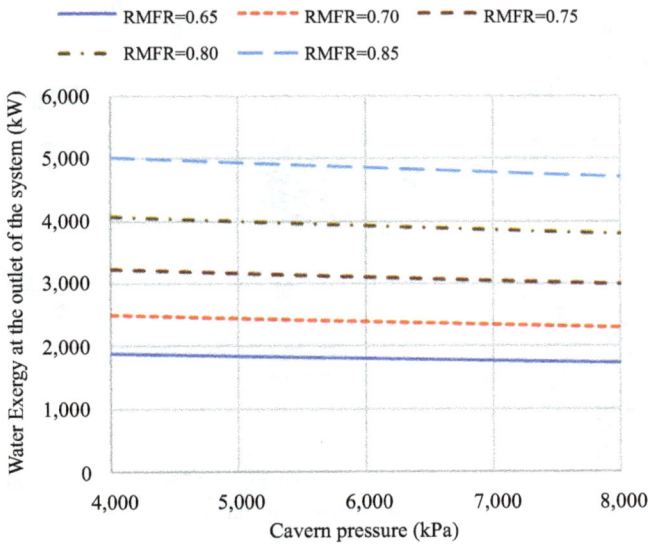

Figure 4.8   Water exergy at the outlet of the system at various RMFR values

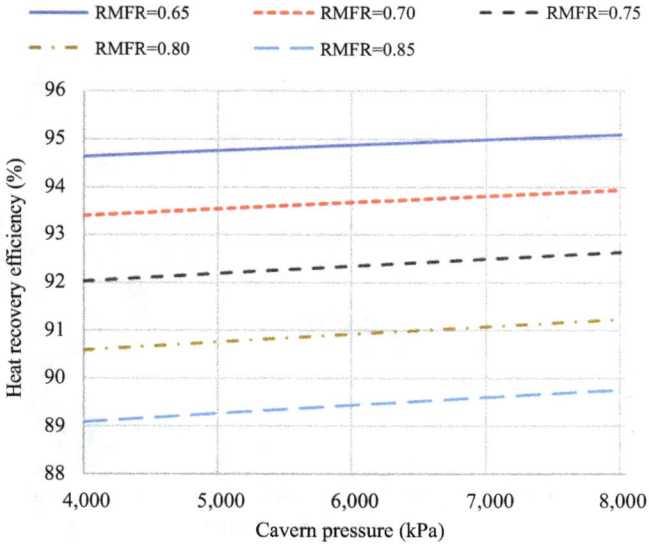

Figure 4.9   Heat efficiency of the discharge cycle at various RMFR for different operating conditions

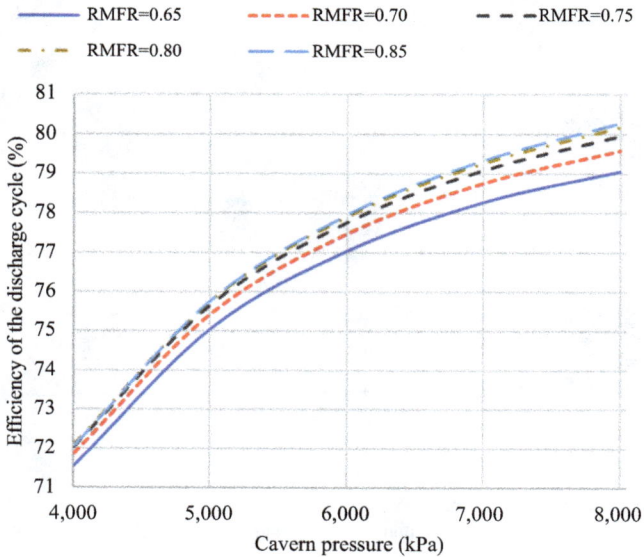

Figure 4.10   Cycle efficiency of the discharge cycle at various RMFR for different operating conditions

Table 4.4   Generated energy by expanders at various operating conditions

Cavern pressure (kPa)	Delivered energy (MWh)				
	RMFR = 0.65	RMFR = 0.70	RMFR = 0.75	RMFR = 0.80	RMFR = 0.85
4,000	399.3	380.7	364.1	349.0	335.2
5,000	436.3	416.0	397.7	381.1	366.0
6,000	462.6	441.0	421.6	403.9	387.9
7,000	482.5	460.0	439.7	421.4	404.6
8,000	498.3	475.1	454.2	435.2	417.9

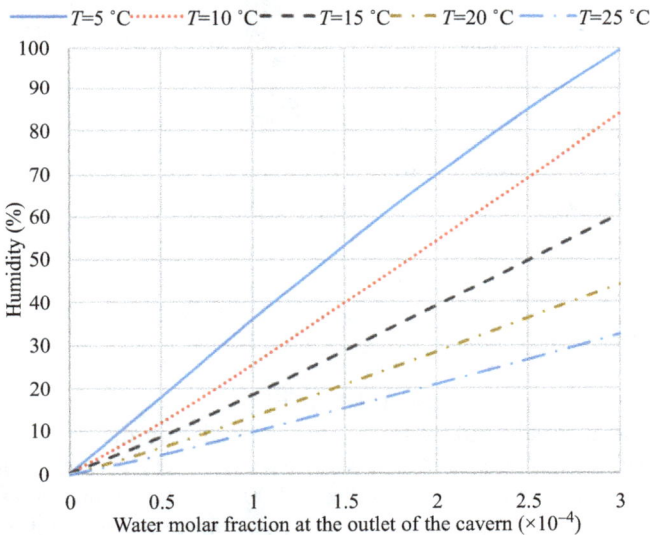

Figure 4.11   Air humidity at various air temperatures and 6,000 kPa

relative humidity conditions for 5 °C air. The corresponding total generated power for RMFR = 0.7 are compared in Figure 4.12. This figure indicates that every 25% increase in relative humidity increases the total expander power by about 0.1%. With the significant follow-on negative impacts associated with increasing humidity in the air stream (corrosion and a requirement for elevated turbine exhaust temperatures to manage dew point), the increase in the output power may not be always economically beneficial. Moreover, there is a cost to air pre-conditioning, so installing equipment to control air humidity should be financially analyzed for each case.

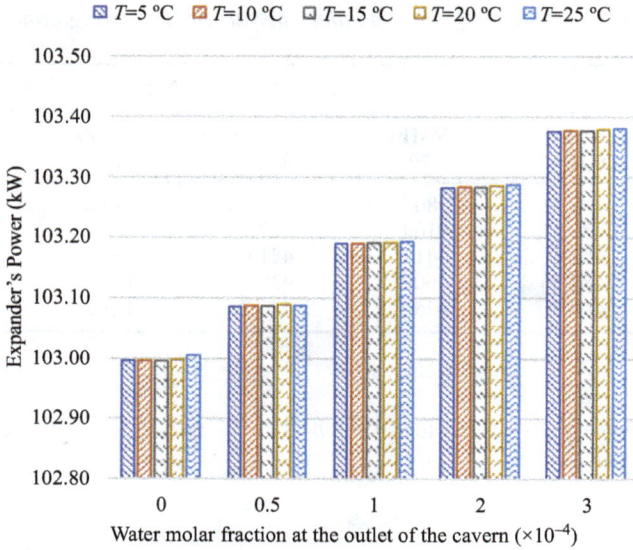

Figure 4.12   *Total generated power by expanders at various water molar fraction in the air stream*

4.5 Conclusion

The objective of this work was to determine the effect of air temperature, air pressure, air to thermal fluid mass rate, and humidity on the performance of the discharge cycle for a 100 MW A-CAES system. In the applied model, it was assumed that the size of water hot tank was 3,000 m^3 with a temperature of 213 °C, and the outlet pressure of the system for all cases was 100 kPa. Simulations were performed for cavern pressures between 4 and 8 MPa, whereas the ratios of hot water mass flow and air mass flow were between 0.65 and 0.85. The obtained result shows that the maximum reachable power and efficiency in the considered ranges were 117 MW and 80.3%, respectively. This study also showed that at lower pressures, the effect of pressure change is more significant than at higher pressures. It was concluded that the maximum efficiency happened at RMFR of about 0.8. In terms of generated energy, a lower air mass flow rate at a higher pressure had a better performance while it was noted that in real problems, cost-benefit analyses are necessary since maintaining the same efficiency in this condition needs more expensive expanders. Furthermore, for every 5 °C increase in the cavern air temperature, the system efficiency increased by about 0.1%. Finally, this work showed that humidity increases the generated power of the system and every 25% increase in humidity percentage can increase the total expanders power by about 0.1%.

References

[1] M. Child, O. Koskinen, L. Linnanen, and C. Breyer, "Sustainability guard-rails for energy scenarios of the global energy transition," *Renewable and Sustainable Energy Reviews*, vol. 91, pp. 321–334, 2018.

[2] O. Dimitriev, T. Yoshida, and H. Sun, "Principles of solar energy storage," *Energy Storage*, vol. 2, no. 1, 2020.

[3] W. He, X. Luo, D. Evans, et al., "Exergy storage of compressed air in cavern and cavern volume estimation of the large-scale compressed air energy storage system," *Applied Energy*, vol. 208, pp. 745–757, 2017.

[4] N. Khan, S. Dilshad, R. Khalid, A.R. Kalair, and N. Abas, "Review of energy storage and transportation of energy," *Energy Storage*, vol. 1, no. 3, 2019.

[5] M. Ebrahimi, R. Carriveau, D.S.-K. Ting, A. McGillas, and D. Young, "Transient thermodynamic assessment of the world's first grid connected UWCAES facility by exergy analysis," in *IEEE*, BREST, France, 2019.

[6] J. Zhan, O.A. Ansari, W. Liu, and C.Y. Chung, "An accurate bilinear cavern model for compressed air energy storage," *Applied Energy*, vol. 242, pp. 752–768, 2019.

[7] X. Luo, J. Wang, M. Dooner, and J. Clarke, "Overview of current development in electrical energy storage technologies and the application potential in power system operation," *Applied Energy*, vol. 137, pp. 511–536, 2015.

[8] N. Hartmann, O. Vöhringer, C. Kruck, and L. Eltrop, "Simulation and analysis of different adiabatic compressed air energy storage plant configurations," *Applied Energy*, vol. 93, pp. 541–548, 2012.

[9] W. Liu, Q.Q Li, F. Liang, L. Liu, G. Xu, and Y. Yang, "Performance analysis of a coal-fired external combustion compressed air energy storage system," *Entropy*, vol. 16, pp. 5935–5953, 2014.

[10] https://hydrostor.ca/, January 15, 2021 [Online].

[11] W. Liu, L. Liu , L. Zhou , et al., "Analysis and optimization of a compressed air energy storage-combined cycle system," *Entropy* , vol. 16, no. 6, pp. 3103–3120, 2014.

[12] Y. Li, X. Wang, D. Li, and Y. Ding, "A trigeneration system based on compressed air and thermal energy storage," *Applied Energy*, vol. 99, pp. 316–323, 2012.

[13] H.S. Yoshida, Y. Oishi, T. Hatoya, R. Echigo, and Y.C. Hang, "Thermofluid behavior in the cavern for the compressed air energy storage gas turbine system," *Heat Transfer, Proceedings of 11th IHTC*, vol. 6, pp. 523–528, 1998.

[14] M. Raju and S.K. Khaitan, "Modeling and simulation of compressed air storage in caverns: A case study of the Huntorf plant," *Applied Energy*, vol. 89, no. 1, pp. 474–481, 2012.

[15] A. Sciacovelli, Y. Li, H. Chen, et al., "Dynamic simulation of adiabatic compressed air energy storage (A-CAES) plant with integrated thermal

storage—Link between components performance and plant performance," *Applied Energy*, vol. 185, pp. 16–28, 2017.

[16] Y. Zhang, K. Yang, X. Li, and J. Xu, "The thermodynamic effect of air storage chamber model on advanced adiabatic compressed air energy storage system," *Renewable Energy*, vol. 57, pp. 469–478, 2013.

[17] Y.S.H. Najjar and N.M. Jubeh, "Comparison of performance of compressed-air energy-storage plant with compressed-air storage with humidification," *Proceedings of the Institution of Mechanical Engineers, Part A: Journal of Power and Energy*, 2006.

[18] M. Ebrahimi, R. Carriveau, D.S-K. Ting, and A. McGillis, "Conventional and advanced exergy analysis of a grid connected underwater compressed air energy storage facility," *Applied Energy*, vol. 242, pp. 1198–1208, 2019.

[19] R. Elarem, T. Alqahtani, S. Mellouli, et al., "A comprehensive review of heat transfer intensification methods for latent heat storage units," *Energy Storage*, vol. 2, pp. 1–30, 2019.

[20] X. Luo, J. Wang, M. Dooner, J. Clarke, and C. Krupke, "Modelling study, efficiency analysis and optimisation of large-scale adiabatic compressed air energy storage systems with low-temperature thermal storage," *Applied Energy*, vol. 162, pp. 589–600, 2016.

[21] M. Ebrahimi, D.S-K. Ting, R. Carriveau, A. McGillis, and D. Young, "Optimization of a cavern-based CAES facility with an efficient adaptive genetic algorithm," *Energy Storage*, vol. 2, no. 6, 2020.

[22] S.D. Garvey and A. Pimm, "Chapter 5: Compressed air energy storage," in *Storing Energy*, Elsevier , 2016, pp. 87–111.

[23] M. Fallah, S.M.S. Mahmoudi, M. Yari, and R. Akbarpour Ghiasi, "Advanced exergy analysis of the Kalina cycle applied for low temperature enhanced geothermal system," *Energy Conversion and Management*, vol. 108, pp. 190–201, 2016.

[24] S. Seyam, I. Dincer, and M. Agelin-Chaab, "Thermodynamic analysis of a hybrid energy system using geothermal and solar energy sources with thermal storage in a residential building," *Energy Storage*, vol. 2, no. 1, 2020.

[25] R. Lanzafame and M. Messina, "A new method for the calculation of gases enthalpy," in *35th Intersociety Energy Conversion Engineering Conference and Exhibit*, 2000.

[26] B. Elmegaard and B. Wiebke, "Proceedings of 24th International Conference on," in *Efficiency of Compressed Air Energy Storage*, Novi Sad, Serbia, 2011.

[27] K. Yang, Y. Zhang, X. Li, and J. Xu, "Theoretical evaluation on the impact of heat exchanger in advanced adiabatic compressed air energy storage system," *Energy Conversion and Management*, vol. 86, pp. 1031–1044, 2014.

[28] R. Carriveau, M. Ebrahimi, D.S-K. Ting and A. McGillis, "Transient thermodynamic modeling of an underwater compressed air energy storage plant: Conventional versus advanced exergy analysis," *Sustainable Energy Technologies and Assessments*, vol. 31, pp. 146–154, 2019.

Chapter 5

Technical feasibility analysis of compressed air energy storage from the perspective of underground reservoir

Li Li[1] and Xiao Lin[2]

Renewable energy production holds promise for meeting the rising energy demand and reducing greenhouse gas emissions. However, renewable energy generation has a negative effect on the stability and flexibility of power systems. To tackle this problem, compressed air energy storage (CAES) technology was proposed. CAES commonly adopts underground caverns as the high-pressure air reservoir, which is a key component of CAES and influences the system performance in many respects. This chapter is intended to discuss the potential and challenges of large-scale commercialization of CAES by analyzing the technical issues associated with the underground reservoir. We also conducted a case study of the CAES in Ontario, Canada, where the salt cavern is used for the energy storage.

Keywords: CAES; underground reservoir; salt cavern; feasibility analysis

5.1 Introduction

Renewable energy refers to the energy generated from natural resources, such as sunlight, wind, tides, geothermal heat, biomass, etc. Renewable energy production holds key potential in addressing global energy shortage and reducing greenhouse gas emissions. Although fossil fuel still dominates global energy consumption (Table 5.1), renewable energy experiences rapid growth, with 41% contribution to the increase in energy consumption. According to the Energy Perspectives report of International Energy Agency (IEA), under the sustainable development scenario, renewable energy will become the major energy recourse, in which wind and solar energy are expected to account for more than half of the total power generation (Figure 5.1). However, the

[1]Department of Civil and Environmental Engineering, University of Waterloo, Waterloo, Ontario, Canada
[2]The York Management School, University of York, York, UK

Table 5.1   Primary energy share and growth in 2019 [2]

Source of energy	Consumption (EJ)	Annual change (EJ)	Primary energy share
Oil	193.0	1.6	33.1%
Gas	141.5	2.8	24.2%
Coal	157.9	−0.9	27.0%
Renewable	29.0	3.2	5.0%
Hydro	37.6	0.3	6.4%
Total	583.9	7.7	

Source: BP Statistical Review of World Energy June 2020.

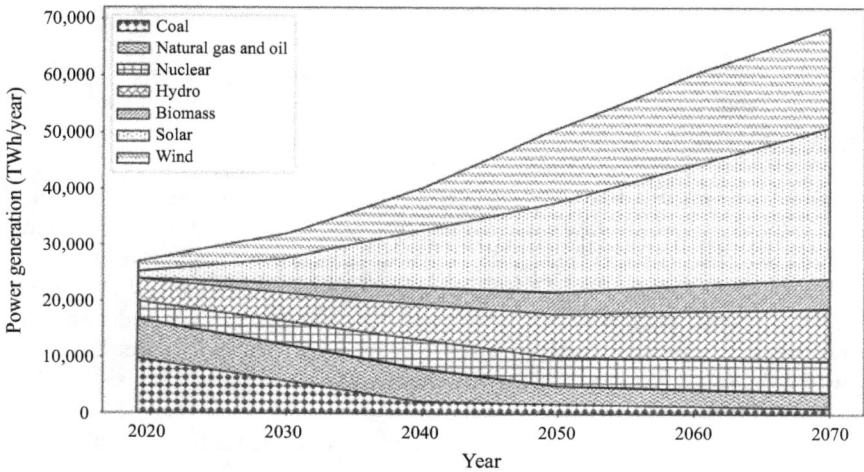

Figure 5.1   Global power generation in the sustainable development scenario [3]

electricity generated from renewable energy suffers from large fluctuation problems and uncertainty risks due to the intermittent and unpredictable natures of energy resources. The energy overproduction during the low-demand period and the insufficient energy generation during peak-demand periods make it difficult to balance energy supply and demand. A possible solution is to combine renewable energy with energy storage technology, which stores energy when demand and generation costs are low or energy sources are intermittent, and then reuse it when demand and generation costs are high or there are no alternative means for power generation [1].

Energy storage is a process of converting one type of energy, which is hard to store, into another form that can be easily stored and converted back to its original form when needed [4]. The energy storage methods can be categorized into five different types: mechanical energy storage, chemical energy storage, electric storage, heat storage, and biological storage. By comparison with other alternatives in terms of technical characteristics, capital energy cost, capital power cost, and cycle efficiency, CAES is regarded as an ideal option for large-scale energy storage, which falls into the category of mechanical energy storage [5].

The fundamental idea of using compressed air as a medium for energy storage dates back to the 1940s [6], but it was not until the 1960s that this technology was applied in the industry. In the 1960s, a large number of nuclear power, lignite coal-fired power plants, and other kinds of plants were constructed, but the power supply was still inadequate during peak time, and on the other hand, a significant amount of energy was wasted in off-peak periods. The electricity price difference between peak and off-peak periods motivated energy storage research. CAES received particular attention because it has the potential to provide quick start-up and grid-scale storage, and to be used in a wide range of operating conditions [7]. In addition, CAES is easier to be integrated with renewable power generation to eliminate the intermittent and fluctuation of electricity supply [8].

5.2   CAES and high-pressure air reservoir

A CAES system comprises a compressor, a turbine, a motor/generator, and an air reservoir. When the electricity load is low, electricity is used to drive compressors to compress air at a higher pressure and store the electricity in the form of internal energy in reservoirs. When the electricity demand is high, the stored high-pressure air in reservoirs is released to drive turbine generators to produce electricity [9]. CAES can be divided into two categories: diabatic CAES and adiabatic CAES. In diabatic CAES, a small amount of natural gas is used during the energy-producing stage to preheat the air before it enters the turbine (Figure 5.2(a)). Although this technology is capable of producing three times more electricity than conventional gas turbines with the same amount of fuel [10], the carbon footprint is still unavoidable. In comparison, the adiabatic CAES technology (Figure 5.2(b)) utilizes thermal energy storage devices to store the heat generated by compression and reuses it for expansion. The adiabatic CAES can increase the efficiency up to 70% and reduce the carbon dioxide emissions [11,12], but further research is needed to mitigate the problems associated with energy storage systems, for example, large energy waste when the air temperature is too high [13].

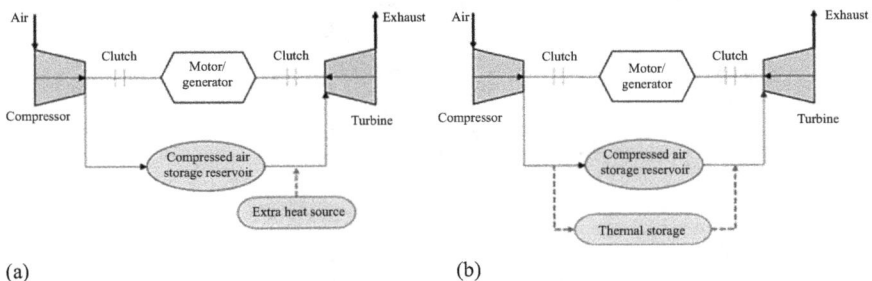

(a)   (b)

Figure 5.2   (a) Diabatic and (b) Adiabatic Compressed Air Storage System [14]

To achieve high efficiency, large CAES power stations introduced multistage compression technique with axial-flow and centrifugal compressors. Multistage expanders are also often employed to drive generators [15]. In addition, careful consideration must be given to the selection and design of compressed air storage reservoirs [16]. Due to the low power and energy density of CAES, a large volume of the reservoir or high-pressure air is required to store large-scale energy. The volume and the pressure tolerance of reservoirs determine the storage capability of the CAES system, and the reliability of the reservoir is crucial in ensuring safe operation. There are two main compressed air storage options: aboveground and underground (Table 5.2).

The aboveground CAES storage media mainly employs pressure vessels, such as high-pressure steel tanks and pipelines, which can withstand high pressure and allow a more flexible layout [19]. The main drawback of aboveground CAES is that constructing high-pressure tanks may be prohibitively expensive, although some advances have been made for storing air in pipeline steel to reduce system cost. The underground storage seems to be more desirable for its low construction cost (Table 5.2). However, one of the predominant challenges of establishing underground CAES is to find geographical formations that are tight enough to prevent the high-pressure air stored in the formations from leaking under cyclic operations. Additionally, the formations should be deep enough to withstand high air pressure [20].

Salt caverns are regarded as an ideal solution for underground CAES storage [21], with at least four prominent advantages [22]. First, salt is easily dissolved in water, which enables a salt cavern to be constructed with solution mining [23] and the cavern shape to be controlled easily [10]. Second, the self-healing capability of salt rocks eliminates air leakage and meantime enhances the operation safety of CAES subject to gas-pressure variation. Third, the permeability of salt rocks is low (10^{-24}–10^{-21} m^2), which ensures that pressured air will not leak from salt caverns. Fourth, it is relatively easier to find salt caverns near renewable-energy production and power-consumption areas because the resource is abundant worldwide. The large mined cavities can be reused as the storage reservoir [24]. The first two CAES

Table 5.2   Cost for different CAES storage media [17,18]

Type	Reservoir	Size (MW)	Power-related plant components cost ($/kW)	Energy storage components cost ($/kWh)	Storage typical hours (h)	Total cost ($/kW)
Aboveground	Surface pipeline	50	350	30	3	440
Underground	Salt	200	350	1	10	360
	Hard rock	200	350	30	10	650
	Porous formation	200	350	0.1	10	351

plants in the world both used salt caverns developed by solution mining, and details will be described later.

Hard rock formations have been used for hydrocarbon storage, e.g., natural gas, for decades due to their airtightness properties and commercially available excavation technologies [25]. Although the output power of the CAES system built in hard rock is higher than salt rocks, the excavation of new hard rock caverns can be costly [26] (Table 5.2). To improve the performance of hard rock CAES systems, hydraulic compensation method is suggested. During the discharging operation, water is injected into the hard rock from surface reservoirs to displace the stored air whereby the air in the hard caverns can stay at a constant pressure to drive the turbine in the CAES plant; during the charging operation, high-pressure gas is injected to displace the water in the cavern. This approach only needs one-fifth of the volume of the salt cavern to reach the same capability of energy storage [27].

Since 1915, porous formations, such as aquifer formations, have been used to store natural gas. Currently, more than 95% of natural gas in storage systems is stored in the porous formation. Although the physicochemical properties and the storage cycles of natural gas are different from those of CAES, most technologies and methodologies used in natural gas storage can be directly applied to CAES, such as site selection, gas compression-system operation, and stability analysis of the reservoirs and systems [16,28–30]. Compared to the salt cavern and hard rock formation, constructing reservoirs in porous formations has the lowest cost. However, the construction of CAES has strict requirements on geological conditions of porous formations. The formations should have sufficient porosity for air storage and high permeability to maintain movement of airflow in the reservoirs during charging and discharging operation. Additionally, the overlying rock layers and adjacent formations must be impermeable and have the structural integrity to prevent air from leaking and escaping to the ground [31]. Besides, some minerals in porous formations may react with the oxygen in the air and produce oxidation products such as gypsum ($CaSO_4 \cdot 2H_2O$), which reduces the porosity of reservoir rocks and decreases the performance of CAES.

5.3 CAES commercial utilization and technical challenges

The commercial CAES projects under construction or planned around the world are listed in Table 5.3. To date, there are only two large CAES plants in commercial operation all over the world: the Huntorf and the McIntosh Plant.

The first CAES facility, Huntorf plant, was built near Bremen, Germany, in 1978 [33]. The plant has been successfully operating for more than 30 years, and is still in excellent operating condition with 99% starting reliability and 90% ability [20]. The CAES plant was initially designed and built to provide the nearby nuclear power units with black-start services and cheap peak power. The ramp-up time for the plant is only 10 minutes. The Huntorf CAES system employs two stages of compressors and expanders to increase efficiency to

Table 5.3   CAES projects all around the world [14]

Name	Country	Power capacity (MW)	High-pressure storage reservoir type	Depth (m)	Cavern volume (m³)	Status
Huntorf	Germany	290	Salt rock cavern	650	310,000	Operation
McIntosh	USA	110	Salt rock cavern	442	580,000	Operation
Wuhu	China	0.5	Pressure vessel	–	–	Operation
Goderich	Canada	2.2	Salt rock cavern	–	–	Operation
Norton	USA	2,700	Hard rock cavern	670	9,600,000	Construction
Iowa Energy Park	USA	270	Porous formation	914	–	Construction
Zhangjiakou	China	100	Air tank	–	–	Construction
Jintan	China	50	Salt rock cavern	–	–	Construction
ADELE	Germany	300	Salt rock cavern	–	–	Planning
Matagorga	USA	540	Salt rock cavern	–	–	Planning
Seneca	USA	150–270	Salt rock cavern	760	150,000	Planning
PG&E	USA	300	Porous formation	–	–	Planning
Datang CAES	China	300	Porous formation	500	900,000	Planning
Feicheng	China	50	Salt rock cavern	–	–	Planning

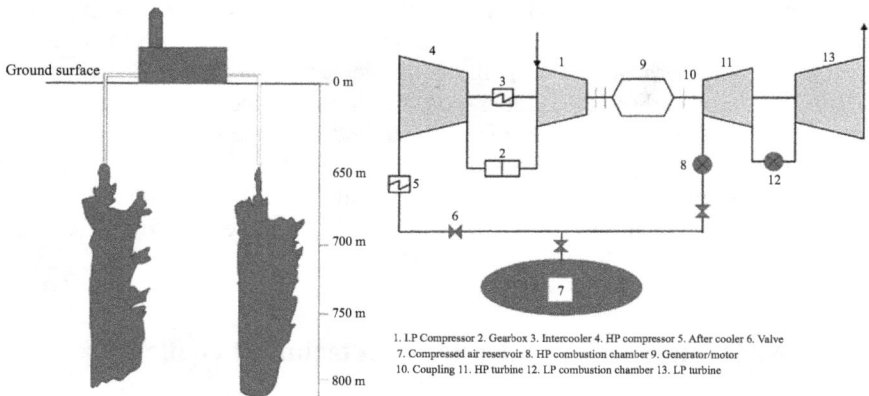

1. LP Compressor 2. Gearbox 3. Intercooler 4. HP compressor 5. After cooler 6. Valve
7. Compressed air reservoir 8. HP combustion chamber 9. Generator/motor
10. Coupling 11. HP turbine 12. LP combustion chamber 13. LP turbine

Figure 5.3   Caverns shape and schematic layout of Huntorf CAES plant [32]

around 42%. Fossil fuel is used as an additional thermal source for heating the compressed air (Figure 5.3). The cavern volume is relatively small (310,000 m³), although two salt caverns are built. The caverns were built in salt formation at depth over 600 m under the ground to store compressed air ranging from 4.8 to 7.0 MPa. Under the working condition of a daily cycle, 290 MW rated power is provided for 2 h after charging for 8 h by injecting compressed air into the salt caverns. Now the plant has been operationally modified to balance wind energy output and is able to supply power for up to 3 h [34].

In 1991, Alabama Electric Cooperative built a CAES plant in southwestern Alabama on the McIntosh salt dome, which is the first CAES facility established in the USA and has started operating commercially since then (Figure 5.4). This plant employs a single cavern that is developed by solution mining. The cavern is located at a depth of 442 m under the ground with the volume of 560,000 m^3. The pressure of the stored compressed air ranges from 4.5 to 7.4 MPa. The loading duration of the plant is up to 26 h. The design of McIntosh is similar to that of the Huntorf CAES plant, but the McIntosh CAES plant made some improvements by storing the heat from the exhausts with a heat recuperator and using it to reheat the air released from the salt cavern to approximately 320 °C. This technology progress reduces about 22% fuel consumption at full-load output and increases the cycle efficiency to 54% [20]. In the early operations, significant outages occurred, but these problems were solved by modifying the mounting of the high-pressure combustor and redesigning the low-pressure combustor. Over 10 years of operations (1998–2008), the McIntosh CAES plant maintained a high average starting reliabilities from 91.2% to 92.1%. The average running reliability for the generation and compression cycle is 96.8% and 99.5%, respectively [35].

Despite the aforementioned successful examples of CAES commercialization, there exist some obstacles that hinder CAES development. For example, the cycle efficiency of commercial CAES plants in operation is more than 50%, and

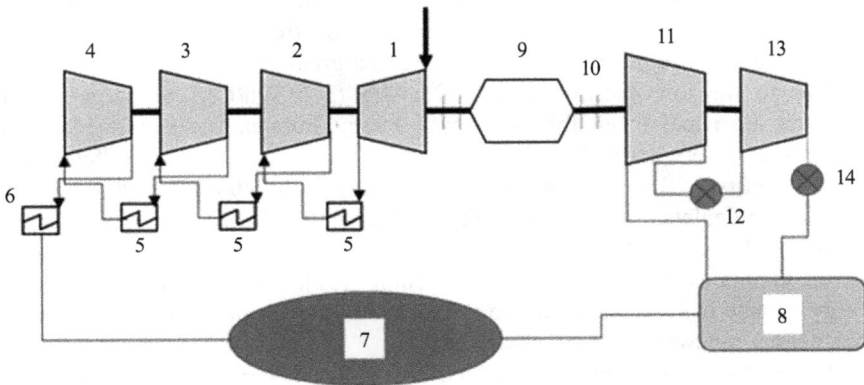

1. LP Compressor 2. IP compressor 1 3. IP compressor 2 4. HP compressor 5. Intercooler 6. Aftercooler 7. Compressed air reservoir 8. Recuperator 9. Generator/motor 10. Coupling 11. LP turbine 12. LP combustion chamber 13. HP turbine 14. HP combustion chamber

Figure 5.4   Schematic layout of the McIntosh CAES plant

Heat storage

Superconducting magnetic energy storage

Electrostatic capacitors

Electrochemical capacitors (supercapacitors)

Flow battery

Molten salt battery

Conventional battery

Flywheel

Compressed air energy storage

Pumped hydroelectric storage

30   40   50   60   70   80   90   100
Cycle efficiency (%)

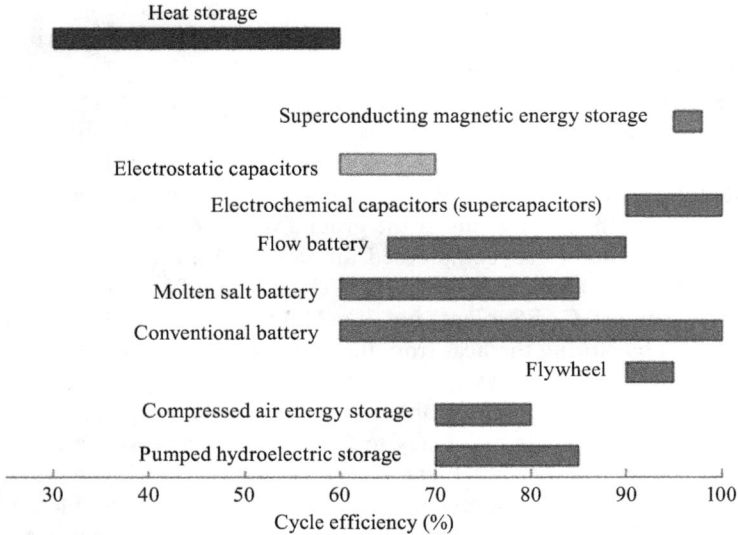

Figure 5.5   Energy storage systems cycle efficiency [5,30]

is expected to reach 70%–80%, which is considerably lower than some other energy storage system (Figure 5.5). Besides, the fluctuation of renewable energy production has high requirements of the dynamic response and robustness of the system under loading scenario. In order to solve these problems, we can optimize thermal storage and exchange systems, improve compressors and turbines, and devise optimal life-cycle control strategies. Aside from that, high-pressure air reservoirs affect the performance of CAES to a great extent.

Due to the low energy density of CAES (3–6 Wh/L) [36], large-volume reservoirs are necessitated for large-scale energy storage. As mentioned above, underground reservoirs (salt cavern, hard rock formation, and porous formation) afford benefits of low construction cost compared to aboveground CAES. In addition, the depleted dissolution mining salt caverns, oil reservoirs, mine tunnels, wellbore hole, and orphan wells can be reused as underground reservoirs, which further facilitates the construction procedure [37]. Nonetheless, underground energy storage has a reliance on the formation with appropriate geological conditions nearby renewable energy resources.

When the underground compressed air storage reservoir is not available in the location of the plant, the pressure vessels become an option with the high cost of building high-pressure vessels. An alternative solution for reducing the storage price and get flexible system arrangement is to liquefy the air and get it stored, which is called liquid air energy storage (LAES) [38]. The LAES has a larger energy density than the conventional CAES system. However, during the cold and heat energy storage process, the energy loss may be very large [14].

5.4   A case study of CAES in Ontario, Canada

Renewable energy is an important energy source in Canada, which occupies more than 18% of the total energy supply. The last decade witnessed tremendous growth in wind and solar energy. The wind electricity generation has experienced an 18-fold increase since 2004, and the data for solar energy are 324-fold (Figure 5.6). As the leader in wind energy of Canada, Ontario has 5,076 MW installed capacity of wind energy, which accounts for 40% of the total installed wind energy in Canada. In addition, Ontario possesses the unique advantages of storing renewable energy to handle the variability of renewable resources. The wind farm and solar farm are mainly located in the southwest of Ontario (Figure 5.7), which has a large amount of salt bed in the Salina Formation. The salt bed forms the basis for constructing salt caverns used as underground reservoirs.

In the Salina Formation, the maximum occurrence thickness of the salt beds is 200 m, making it possible to build salt caverns in this area for large-scale energy storage. The soluble component in Ontario's salt rock is up to 98%, which is amenable to the application of solution mining. In addition, solution mining was carried out in this area for decades since 1960, and the existing salt caverns provide an economically attractive option for building CAES facilities [42].

The main Salina Formation in Ontario can be divided into two types: upper Salina and lower Salina, as illustrated in Figure 5.8. The units containing salt in upper Salina are F, D, and B. Although the rock salt is pure, some interbedded layers of the salt rocks are present. In the F unit, the beds of shale appear between the layers of the salt in addition to the shaly dolomite and fine crystalline buff. The salt in D unit is nearly pure but divided by a thin layer of buff dolomite. B unit is

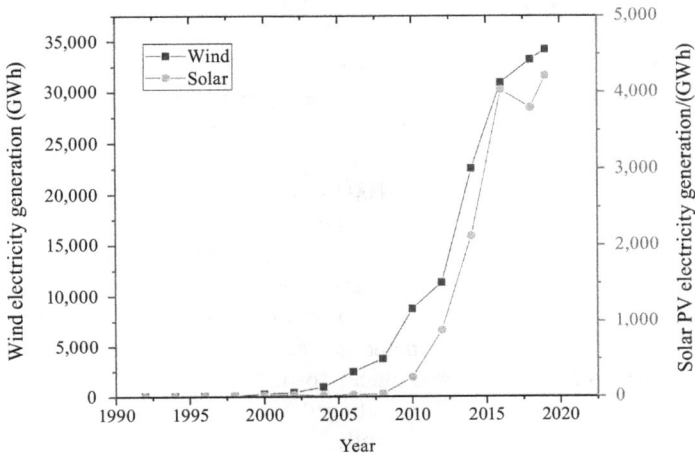

Figure 5.6   Wind and solar energy generation in Canada [39]

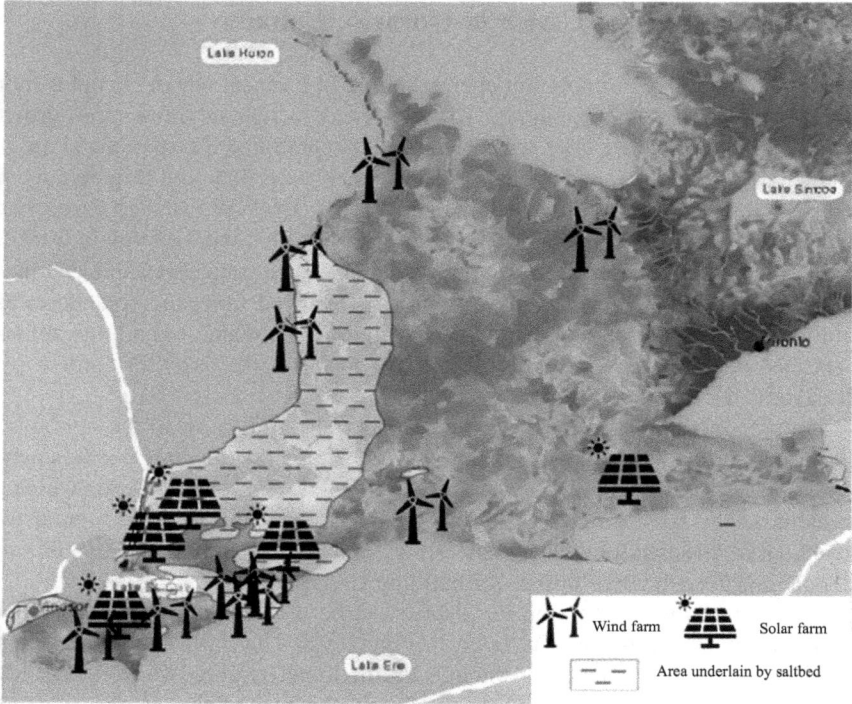

*Figure 5.7   Wind, solar energy resources, and salt bed distribution in Ontario
[40,41]*

the main salt unit in upper Salina. It is the thickest salt layer among the four units
with a thickness of about 90 m, but also includes some thin dolomite layers. Unit
A2 is the only unit containing salt in the lower Salina. The thickness of the salt
layer in A2 is about 45 m. It is interbedded with several kinds of dolomite.

Considering the presence of thin insoluble layers in Ontario Salina Formation, it
is challenging to construct vertical salt caverns in Ontario like that built in Huntorf
and McIntosh. To overcome this difficulty, we can opt for horizontal caverns.
However, the large roof span (e.g., 75–100 m) of horizontal salt caverns in a thin salt
bed may incur the instability of caverns. Therefore, appropriate measures should be
taken to control the dissolution process and the shape and volume of horizontal salt
caverns, but the research in this area has rarely been reported. In the future work, it is
worth investigating the dissolution patterns of salt rock during the construction of
horizontal caverns and developing methods for optimizing the dissolution process.

The renewal resource and geological formation in Ontario are amenable to
CAES construction. However, when setting the CAES plant, we still need to take
into account the following factors carefully. First, the seismic or earthquake would
be potential sources of hazards for the operation of CAES plant with high-pressure
reservoirs in geological formations, so the plant site should be located in an area

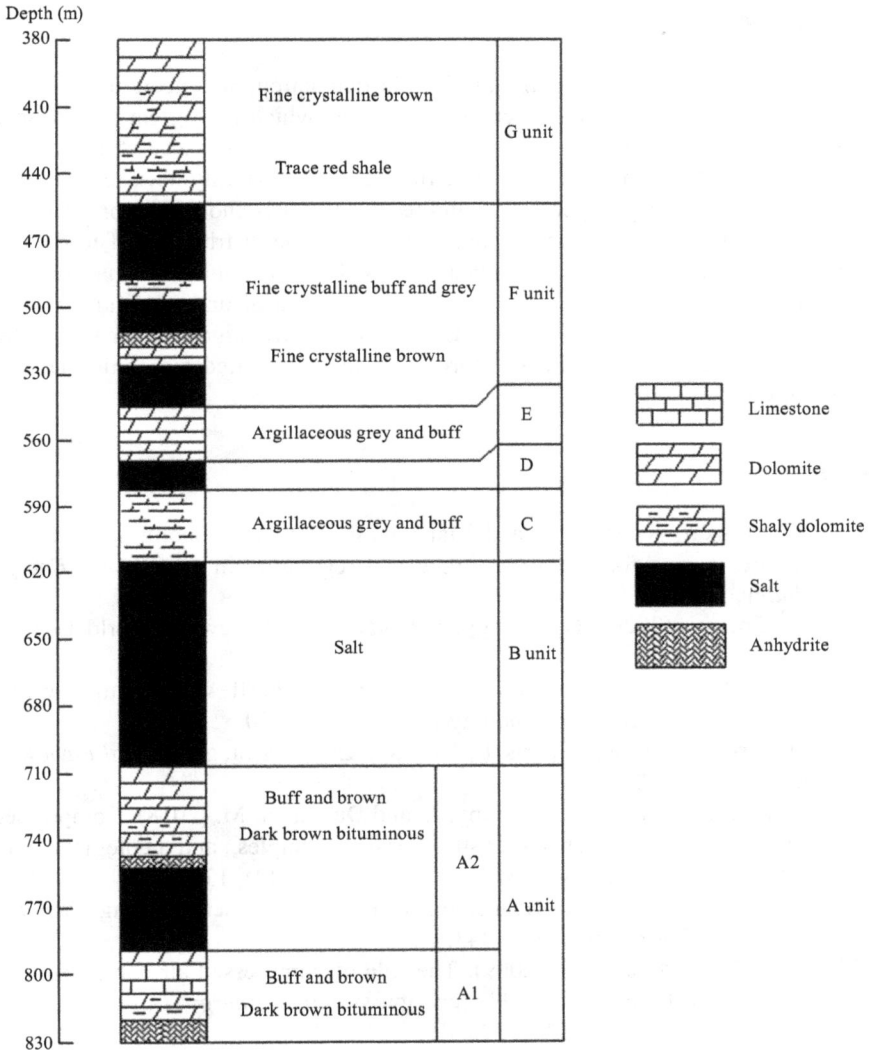

Figure 5.8   The Salina Formation Subdivision [43]

without minimal seismic or earthquake fault problems. Second, because of the high-pressure air issues in the CAES plant, the site should be away from other hazardous facilities. Third, the daily operation of the CAES plants gives rise to noise pollution; in this case, the site of the CAES should avoid the high population density area. Last but not least, although the major part of the CAES plant is the underground reservoirs, the facilities above ground still require a footprint for operation. For example, the McIntosh plant occupies 10 acres for the plant facilities.

5.5   Conclusion

CAES provides an effective method to handle fluctuation and uncertainty of electricity generated from renewable energy resources, which is of great significance for energy security and sustainability transition. In spite of the successful examples of commercial utilization of CAES, its efficiency and flexibility need to be further improved. To this end, the geological and technical issues should be considered in the design and operation of underground reservoir. The distribution of renewable resources and salt rock resources in Ontario is used as a case study for analyzing the feasibility of CAES development from the perspective of underground reservoir construction. The site selection of the CAES plant is investigated after the geological study on the underground salt formation and the related dissolution control technology.

References

[1]   Walawalkar, R., Apt, J., and Mancini, R. (2007). Economics of electric energy storage for energy arbitrage and regulation in New York. *Energy Policy*, 35(4), 2558–2568.

[2]   BP Energy. (2020). BP Energy: BP Statistical Review of World Energy, June 2020.

[3]   IEA (2020). Energy Technology Perspectives 2020, IEA, Paris https://www.iea.org/reports/energy-technology-perspectives-2020.

[4]   McLarnon, F. R., and Cairns, E. J. (1989). Energy storage. *Annual Review of Energy*, 14(1), 241–271.

[5]   Li, L., Liang, W., Lian, H., Yang, J., and Dusseault, M. (2018). Compressed air energy storage: characteristics, basic principles, and geological considerations. *Advances in Geo-Energy Research*, 2(2), 135–147.

[6]   Kalhammer, F. R., and Schneider, T. R. (1976). Energy storage. *Annual Review of Energy*, 1(1), 311–343.

[7]   Lund, H., and Salgi, G. (2009). The role of compressed air energy storage (CAES) in future sustainable energy systems. *Energy Conversion and Management*, 50(5), 1172–1179.

[8]   Cazzaniga, R., Cicu, M., Rosa-Clot, M., Rosa-Clot, P., Tina, G. M., and Ventura, C. (2017). Compressed air energy storage integrated with floating photovoltaic plant. *Journal of Energy Storage*, 13, 48–57.

[9]   Hadjipaschalis, I., Poullikkas, A., and Efthimiou, V. (2009). Overview of current and future energy storage technologies for electric power applications. *Renewable and Sustainable Energy Reviews*, 13(6–7), 1513–1522.

[10] Connolly, D. (2009). A review of energy storage technologies: for the integration of fluctuating renewable energy. Tech. rep, University of Limerick, Limerick, Ireland.

[11] Sciacovelli, A., Li, Y., Chen, H., et al. (2017). Dynamic simulation of adiabatic compressed air energy storage (A-CAES) plant with integrated

thermal storage—link between components performance and plant performance. *Applied Energy*, 185, 16–28.

[12]   Wang, J., Ma, L., Lu, K., Miao, S., Wang, D., and Wang, J. (2017). Current research and development trend of compressed air energy storage. *Systems Science & Control Engineering*, 5(1), 434–448.

[13]   Liu, J. L., and Wang, J. H. (2016). A comparative research of two adiabatic compressed air energy storage systems. *Energy Conversion and Management*, 108, 566–578.

[14]   Tong, Z., Cheng, Z., and Tong, S. (2020). A review on the development of compressed air energy storage in China: technical and economic challenges to commercialization. *Renewable and Sustainable Energy Reviews*, 135, 110178.

[15]   Haisheng. C., Jinchao, L., Huan G., Yujie, X., and Chunqing, T., (2013) Technical principle of compressed air energy storage system. *Energy Storage Science Technology*, 2(02), 146–151.

[16]   Barnes, F. S., and Levine, J. G. (Eds.). (2011). *Large Energy Storage Systems Handbook*. Boca Raton, FL: CRC Press.

[17]   Zakeri, B., and Syri, S. (2016). Corrigendum to "Electrical energy storage systems: a comparative life cycle cost analysis" [*Renewable and Sustainable Energy Reviews* 42 (2015) 569–596]. *Renewable and Sustainable Energy Reviews*, 100(53), 1634–1635.

[18]   Balitskiy, S., Bilan, Y., Strielkowski, W., and Štreimikienė, D. (2016). Energy efficiency and natural gas consumption in the context of economic development in the European Union. *Renewable and Sustainable Energy Reviews*, 55, 156–168.

[19]   Jewitt, J. Mini-Compressed Air Energy Storage for Transmission Congestion Relief and Wind Shaping Applications (Prepared for New York State Energy Research and Development Authority).

[20]   Luo, X., Wang, J., Dooner, M., Clarke, J., and Krupke, C. (2014). Overview of current development in compressed air energy storage technology. *Energy Procedia*, 62, 603–611.

[21]   Li, L., Gracie, R., and Dusseault, M. B. (2020, September). Numerical simulation of salt rock dissolution. In *54th US Rock Mechanics/ Geomechanics Symposium*. American Rock Mechanics Association.

[22]   Wang, T., Yan, X., Yang, H., Yang, X., Jiang, T., and Zhao, S. (2013). A new shape design method of salt cavern used as underground gas storage. *Applied Energy*, 104, 50–61.

[23]   Reda, D. C., and Russo, A. J. (1986). Experimental studies of salt-cavity leaching by freshwater injection. *SPE Production Engineering*, 1(01), 82–86.

[24]   Swift, G. M., and Reddish, D. J. (2005). Underground excavations in rock salt. *Geotechnical & Geological Engineering*, 23(1), 17–42.

[25]   Zhu, H., Chen, X., Cai, Y., Chen, J., and Wang, Z. (2015). The fracture influence on the energy loss of compressed air energy storage in hard rock. *Mathematical Problems in Engineering*, 2015.

[26]   Succar, S., and Williams, R. H. (2008). Compressed air energy storage: theory, resources, and applications for wind power. *Princeton Environmental Institute Report*, 8, 81.

[27]   Schainker, R. B., and Nakhamkin, M. (1985). Compressed-air energy storage (CAES): overview, performance and cost data for 25MW to 220MW plants. *IEEE Transactions on Power Apparatus and Systems*, (4), 790–795.

[28]   Buschbach, T. C., and Bond, D. C. (1974). Underground storage of natural gas in Illinois, 1973. Illinois Petroleum no. 101.

[29]   Evans, D. J., and West, J. M. (2008). An appraisal of underground gas storage technologies and incidents, for the development of risk assessment methodology. *Research Rep.* RR605.

[30]   Ibrahim, H., Ilinca, A., and Perron, J. (2008). Energy storage systems—characteristics and comparisons. *Renewable and Sustainable Energy Reviews*, 12(5), 1221–1250.

[31]   Eckroad, S., and Gyuk, I. (2003). *EPRI-DOE Handbook of Energy Storage for Transmission & Distribution Applications*. Electric Power Research Institute, *Inc*, 3–35.

[32]   Réveillère, A., and Londe, L. (2017, September). Compressed air energy storage: a new beginning. In *Solution Mining Research Institute Fall 2017 Technical Conference*, Munster, Germany.

[33]   Crotogino, F., Mohmeyer, K. U., and Scharf, R. (2001). Huntorf CAES: more than 20 years of successful operation, *SMRI Spring Meeting*; 2001, Orlando, FL, 351–357.

[34]   Van der Linden, S. (2006). Bulk energy storage potential in the USA, current developments and future prospects. *Energy*, 31(15), 3446–3457.

[35]   Biasi, V. D. (1998). 110 MW McIntosh CAES plant over 90% availability and 95% reliability. *Gas Turbine World*, 28, 26–28.

[36]   Chen, H., Cong, T. N., Yang, W., Tan, C., Li, Y., and Ding, Y. (2009). Progress in electrical energy storage system: a critical review. *Progress in Natural Science*, 19(3), 291–312.

[37]   Alexander, S., Cai, J., Flinkfelt, L., and Li, L. (2020). Upcycling Orphan Wells in Alberta: repurposing opportunities using a new evaluation system. *GeoConvention* 2020.

[38]   Peng, X., She, X., Cong, L., et al. (2018). Thermodynamic study on the effect of cold and heat recovery on performance of liquid air energy storage. *Applied Energy*, 221, 86–99.

[39]   IEA (2020), Renewables 2020, IEA, Paris https://www.iea.org/reports/renewables-2020.

[40]   Konrad, J., Carriveau, R., Davison, M., Simpson, F., and Ting, D. S-K. (2012). Geological compressed air energy storage as an enabling technology for renewable energy in Ontario, Canada. *International Journal of Environmental Studies*, 69(2), 350–359.

[41]   Hewitt, D. F. (1962). Salt in Ontario. Ministry of Natural Resources, Division of Mines.

[42] Amiryar, M. E., and Pullen, K. R. (2017). A review of flywheel energy storage system technologies and their applications. *Applied Sciences*, 7(3), 286.

[43] Frizzell, R., Cotesta, L., and Usher, S. (2011). Regional Geology-Southern Ontario. Tiverton, OPG's Deep Geol. Repos. Low Intermed. Level Waste.

Chapter 6

Comprehensive overview of compressed air energy storage systems

Marcos A. Salvador[1], Lenon Schmitz[2], Telles B. Lazzarin[2] and Roberto F. Coelho[2]

Compressed air energy storage (CAES) is a technology employed for decades to store electrical energy, mainly on large-scale systems, whose advances have been based on improvements in thermal management of air compression and expansion stages through adiabatic and nearly isothermal processes. Recently, small-scale CAES (SS-CAES) systems have also been applied as an alternative to replace batteries in autonomous systems and in distributed generation applications with renewable sources. These systems require compact and efficient power stages, with remarkable presence of power electronics. In this context, this chapter presents a comprehensive overview about some CAES and SS-CAES systems and describes their operating principles, as well as information regarding energy density, efficiency, cost, limitations, and challenges to be overcome in order to make them attractive solutions.

Keywords: Compressed air energy storage; hybrid systems; power electronics; small-scale CAES

6.1 Introduction

The current technological development and the growing concern for the environment have contributed to a noteworthy increase in the exploitation of renewable generation sources. However, the variable and intermittent nature of these power sources, such as wind and photovoltaic, presents a major challenge to their extensive penetration into the grid [1–3]. The problems associated with the peak demand and the power system stability caused by renewable sources can be minimized through the application of energy storage systems next to the power plants, in

[1]Federal Institute of Santa Catarina, Department of Electrical Engineering, Jaragua do Sul, SC, Brazil
[2]Federal University of Santa Catarina, Department of Electrical and Electronics Engineering, Florianopolis, SC, Brazil

support to the transmission network, at various points in the distribution grid, and also on the consumer side [2].

The literature describes different ways of performing energy storage, with emphasis on the use of battery banks and supercapacitors of different technologies, as well as the usage of fuel cells, flywheels, pumped-storage hydroelectricity, and compressed air, among others [1]. Particularly, CAES systems store energy in the air compression process, to posterior use during its expansion, being normally divided into small-scale (≤ 100 kWh) and large-scale (>100 kWh) systems.

Large-scale CAES systems have been used to support the electric power system (EPS) since the 1970s. Basically, they are installed to store energy in large reservoirs during periods of low energy demand (off-peak) for later use during higher-demand (peak load) periods [4]. Currently, CAES systems continue to be the target of several studies, but the focus is on their application to reduce power fluctuations due to the intermittency of generation caused by the high penetration of renewable sources in the EPS [5,6]. Over the years, different strategies related to CAES systems have been addressed in order to increase their efficiency and minimize their environmental impacts, as the systems that reuse the heat generated by the air compression process to reduce fuel consumption during its expansion [7].

SS-CAES systems are quoted as alternatives for applications that generally use batteries as storage energy devices. SS-CAES has some advantages from the ecological point of view and life span when compared to commercial batteries. Nonetheless, they also present challenges to be overcome, specially related to their lower energy density and performance. These factors have stimulated the search for methods to increase the efficiency in the air expansion process and the development of strategies for tracking their maximum power or maximum efficiency points [8–10]. In addition, since they are mechanical systems and therefore present a slow dynamic response to transients, hybrid combinations of SS-CAES with other storage devices, such as batteries and supercapacitors, have also been investigated [8,11].

Taking it into account, the evolution of SS-CAES can make it an option among high-density energy storage systems. For example, it may be applied in areas of power electronics already consolidated, such as uninterruptible power supplies, or in emerging areas, like isolated hybrid systems, active networks, and distributed generation systems, solving problems related to intermittent generation, load leveling, and peak shaving. Currently, all of these applications employ batteries; however, solutions are still being sought in other forms of storage that are ecologically less aggressive and have a longer life span.

In light of this, the present chapter offers a review regarding energy storage systems in the form of compressed air, with the aim of pointing out the various existing approaches and highlighting the efforts made in search of new strategies for these types of storage systems. Furthermore, this chapter demonstrates the potential and challenges of using power electronics applied to power processing in CAES systems.

6.2   Compressed air energy storage

Atmospheric air consists of a colorless, odorless, and tasteless mixture of gases, in the approximate proportion of 78% nitrogen, 21% oxygen, and 1% other elements. Since it is a gas, it has the property of compression and can be stored in reservoirs. In this condition, there is an increase in the number of air molecules per unit volume and, consequently, an increase in the internal pressure of the reservoir. The air compression requires the usage of an external source of energy, part of which is stored and returned to the system during the expansion process [12].

Large-scale CAES systems applied to the electrical utility have two main processes: compression and expansion. Basically, the compressed air generated is stored in large underground reservoirs, as salt and rock caverns or depleted gas fields, in order to be later used to generate energy from its expansion [4,5], similarly to a conventional gas turbine [13]. The compression of air creates heat, while the expansion removes heat. If the heat generated during compression can be stored and used during expansion, the efficiency of the storage improves considerably [7]. Depending on how the heat is managed during the compression, CAES technologies can be differentiated into diabatic, adiabatic, and isothermal [14]. The diabatic CAES (D-CAES), also known as the traditional technology, is employed commercially since the 1970s. In this type of system, the heat resulting from air compression is wasted to the ambient, resulting in low efficiency. To overcome this issue, the adiabatic and isothermal approaches have been proposed, including intermediate steps for managing heat exchanges. The adiabatic CAES (A-CAES) consists of a process for conserving and reusing the heat generated during the compression. Conversely, in the isothermal CAES (I-CAES) solution, the fluid temperature is kept approximately constant throughout the process [5,7]. Further information and the technological implementation of these types of CAES are presented in the following.

6.2.1   D-CAES systems

The D-CAES systems can be divided into two distinct generations. In the first-generation D-CAES system, the heat resulting from the compression process is transferred to the atmosphere and the air is reheated during expansion, by burning some fuel, such as natural gas or oil. The second-generation D-CAES system works in a similar way, but the compression and expansion processes are divided into stages, and part of the thermal energy contained in the air that leaves the turbine is used to preheat the air leaving the reservoir [5]. The D-CAES system with the inclusion of a heat exchanger is illustrated in Figure 6.1. This system has been implemented in the McIntosh plant, installed in the USA, in 1991, with a reduction of approximately 22% in fuel consumption, when compared to the first-generation Huntorf plant, installed in Germany, in 1978.

Even with the inclusion of the heat exchanger, this system burns a large amount of fuel, which introduces negative aspects from the point of view of CO_2 emissions, in addition to penalizing its overall efficiency. In general, the efficiency

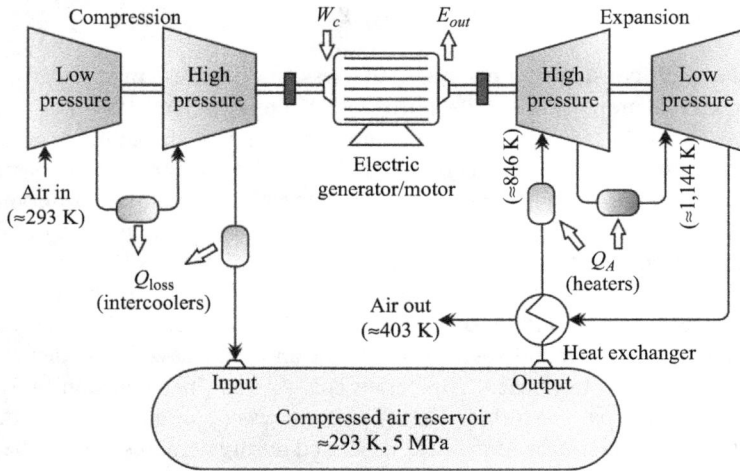

Figure 6.1   Second-generation D-CAES system overview [7]

of this system can reach 54%, being calculated from a relationship between the output energy and the sum of the electrical energy, used for compressing the air, and the energy added to the process in the form of heat [7,15].

6.2.2   A-CAES systems

The A-CAES system has an arrangement that stores the heat generated during the air compression, making it available for the expansion process and, therefore, reducing or even eliminating the fuel consumption. The main benefits of this type of system are the increase in efficiency and the reduction of carbon emissions due to the minimization of fuel use [5,7]. Figure 6.2 depicts the system designed in the advanced A-CAES (AA-CAES) project [16] with its heat exchangers and thermal energy storage, yielding efficiencies of up to 70%.

In this system, the use of heat is carried out by means of exchangers installed close to the air compression and expansion processes. The storage of heat, in turn, can be done in a liquid medium with materials that do not change their state for the given temperature range [7]. Currently, the first AA-CAES demonstration plant, named as ADELE, with storage capacity of 360 MWh, is under development in Germany [17]. Other pilot-scale plants have also been implemented [18].

6.2.3   I-CAES systems

The I-CAES system is an alternative option that eliminates the need for thermal storage at high temperatures and for burning fuel in the process. The I-CAES system tries to approach the ideal isothermal compression/expansion process, keeping the temperature constant throughout the process by means of efficient heat exchange methods, resulting in efficiencies close to 75% [15].

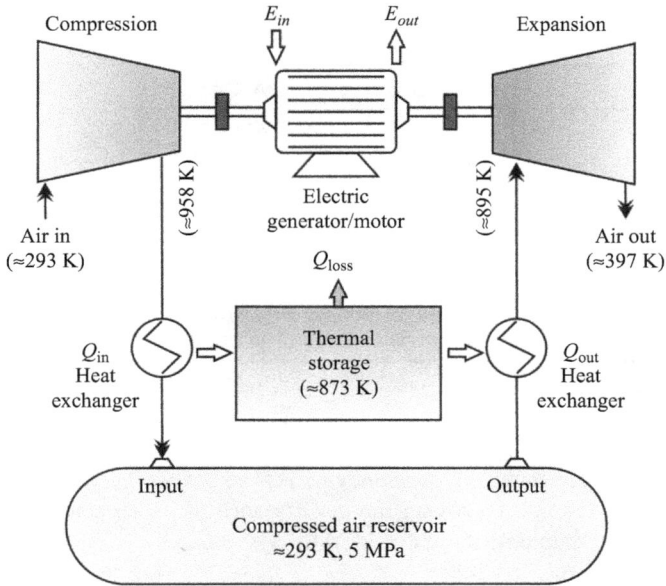

Figure 6.2   AA-CAES system overview [7]

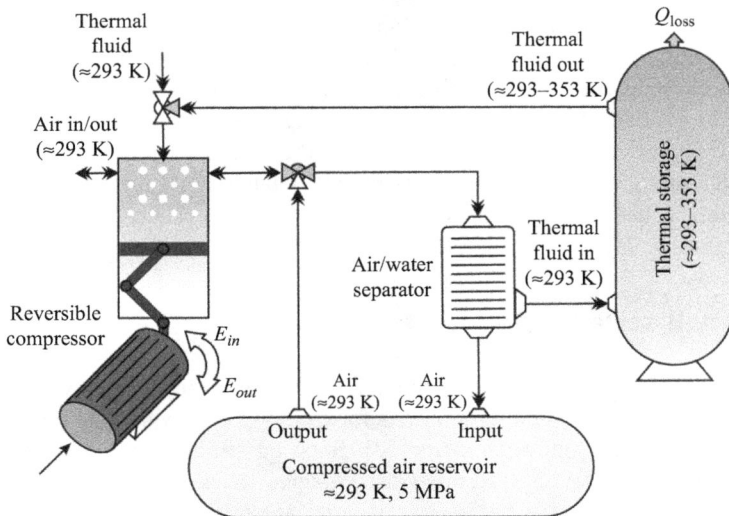

Figure 6.3   I-CAES system overview [7]

The operation of the I-CAES system involves the following steps: the injection of liquid (water/oil) in a plunger cylinder during air compression, the separation of compressed air and liquid, and finally the injection of hot liquid inside the cylinder during expansion [7]. Figure 6.3 illustrates the I-CAES system with the respective

Table 6.1   Comparison among the D-CAES, A-CAES, and I-CAES [7]

	D-CAES	A-CAES	I-CAES
System efficiency	42%–54%	65%–70%	70%–80%
System pressure	4–8 MPa	7–20 MPa	5–40 MPa
Energy density	5 kWh/m^3 at 8 MPa	20 kWh/m^3 at 20 MPa	53 kWh/m^3 at 40 MPa
Temperature	up to 1,023 K	773–873 K	less than 353 K
Burning fuel	Natural gas	Minimal	None
Cost per capacity	760–1,200 US$/kW	850–1,870 US$/kW	1,500–6,000 US$/kW
Technical maturity	Commercial	In project	Pilot projects

Source: Adapted from [7].

described steps. Increasing the efficiency of heat exchange between air and liquid for high flow levels and improving the performance of its separators are the main challenges for the implementation of I-CAES systems.

6.2.4   Summary

Table 6.1 shows a comparative summary of the three primary CAES systems. The figures of merit listed in Table 6.1 show the superiority of the A-CAES and I-CAES systems when compared to the traditional D-CAES, especially from the point of view of efficiency and environmental impacts due to CO_2 emissions. Another advantage to be highlighted is their higher energy densities because of the higher-pressure level used, due to the technological development of compression techniques. Although the estimated capital costs for the I-CAES systems are higher than its counterparts [7], such difference is mainly related to its application on pilot-scale projects, considering, for example, the usage of pressure vessels instead of underground geological reservoirs [19,20].

6.3   Small-scale CAES

Power electronics are often used as an interface between the storage system and the power grid, enabling several possibilities for applying energy storage together with the concept of distributed generation [2]. Such possibilities have stimulated the development of SS-CAES systems with strategies for maximum efficiency point tracking (MEPT) [8] or for maximum power point tracking (MPPT) [9], since both power and efficiency of SS-CAES systems vary depending on factors such as pressure, temperature, and flow.

These applications have also led to the development of hybrid solutions with distinct strategies. Among the proposed hybrid SS-CAES systems, one can mention the employment of compressed air and supercapacitors energy storage (CASCES), and the usage of battery with oil-hydraulics pneumatics (BOP) of types A and B.

Basically, type-A BOP systems have compression/expansion cycles with sealed gas, whereas type-B BOP systems present near-isothermal compression/expansion processes [8].

The SS-CAES and hybrid SS-CAES solutions consist in structures based on power converters; however, studies that address converter topologies, modeling, control strategies, and integration of these systems are still restricted. Therefore, taking as a reference the storage systems based on the use of batteries, there is a need to detail the features of the SS-CAES and hybrid SS-CAES solutions to advance the development of this area.

6.3.1 SS-CAES systems

The SS-CAES systems have been studied as an alternative to replace batteries in uninterruptible power supply, autonomous systems, and in distributed generation applications [7,9]. Figure 6.4 shows one of these small-scale systems, where the atmospheric air is typically compressed by a volumetric compressor and stored under pressure in open-air tanks. When released into the atmosphere, the stored air is used to move a pneumatic motor or a microturbine which drives a dc generator. The generated energy is thus injected into the power grid or applied to a remote load by means of power converters.

SS-CAES systems do not burn fuel as in some large-scale CAES systems, causing fewer negative impacts on the environment when compared to electrochemical batteries, which generate toxic waste and have less longevity. However, the energy density and efficiency of SS-CAES systems are low, implying a greater volume to supply the same amount of energy as batteries [7–9].

As can be seen in Figure 6.4, a control system regulates the air discharge valve and provides parameters to the power converter based on the measurement of some

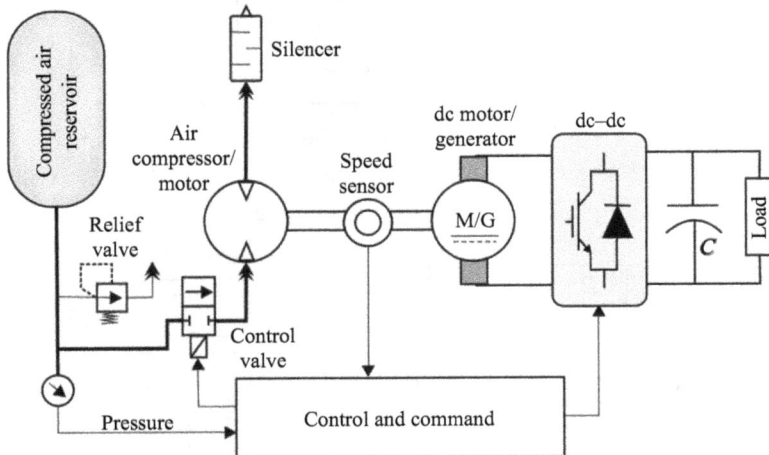

Figure 6.4 SS-CAES system powering remote load [8]

quantities, such as system pressure and generator shaft speed. In general, energy processing and storage systems look for methods to reduce the losses in order to achieve the maximum possible efficiency. In this respect, the studies targeted to SS-CAES systems have addressed strategies for the charging and discharging processes of the compressed air tank, as well as their application associated with photovoltaic and wind generators.

In [21], an air pump based on an air/liquid piston is proposed. This small-scale compressor has a low compression rate and low power, but it can be installed with individual photovoltaic modules in residential applications, whose output power is typically below 300 W. Conversely, in [9], a control strategy for the discharge of compressed air from an SS-CAES system has been designed and implemented by tracking the maximum power point of a pneumatic vane air motor. As illustrated in Figure 6.5, the system drives a permanent magnet dc generator which feeds a resistive load through a buck converter.

This system has been analyzed using a small-signal model and employing the perturb-and-observe method with small speed steps to seek convergence. The proposed control system does not need to monitor the pressure and flow of the compressed air; it only analyses the generator shaft speed and the output current of the buck converter to track the maximum power line shown in Figure 6.6 [9]. From the equations of torque, airflow, and motor speed, it is possible to trace the surface correlated to the magnitudes of power, pressure, and speed of the air motor, and to establish the maximum power line. According to Figure 6.6, the maximum power line depends on the supply pressure and the motor speed [9].

By contrast, in [8], the application of a 100 W pneumatic vane motor for pneumatic/mechanical conversion has been investigated and a strategy for tracking

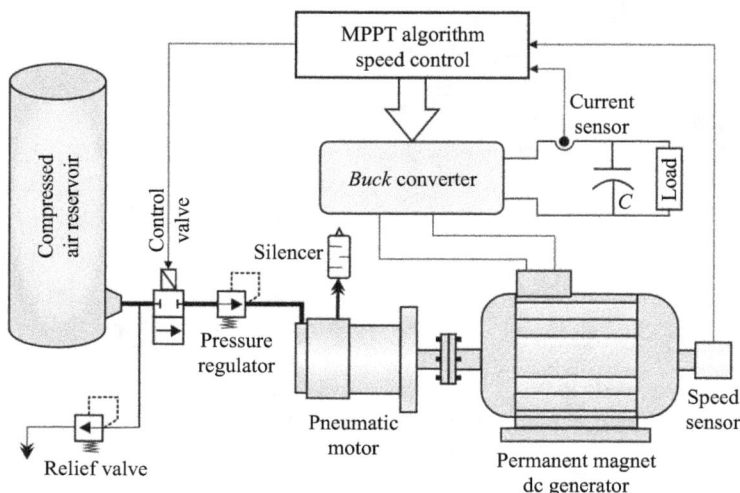

Figure 6.5   Process of air discharge with MPPT in SS-CAES [9]

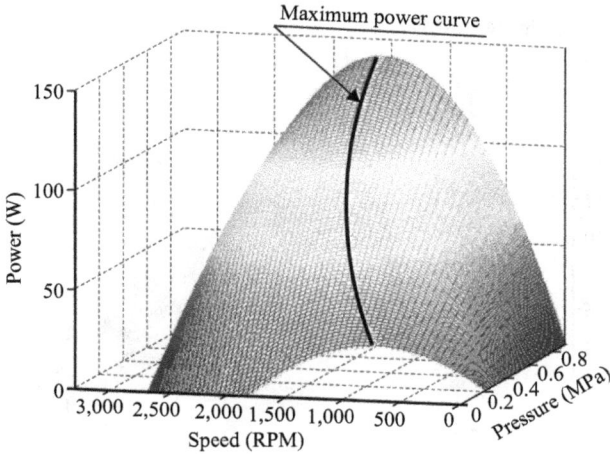

Figure 6.6 Pneumatic motor maximum power line [9]

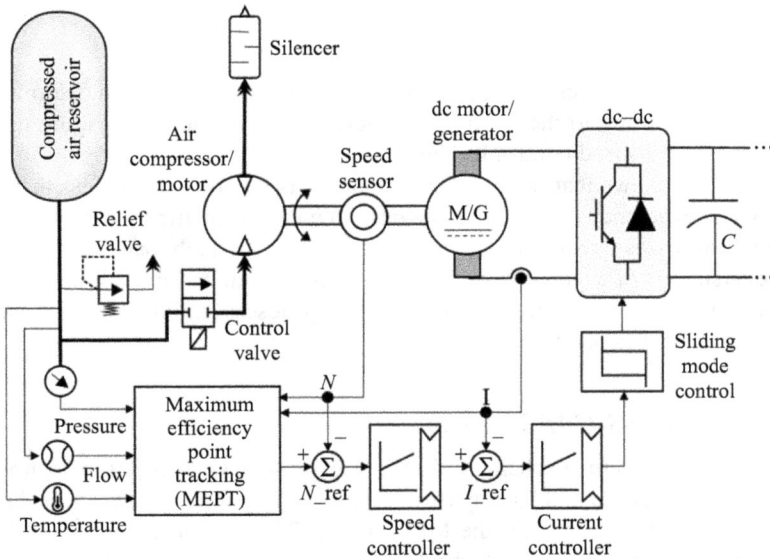

Figure 6.7 SS-CAES with MEPT strategy [8]

its maximum efficiency point is developed according to Figure 6.7. The pressure and load variations are related to the airflow and the speed of the pneumatic air motor, which directly affect its performance. The principle of the MEPT strategy of Figure 6.7 consists in optimizing the energy conversion based on the measurement

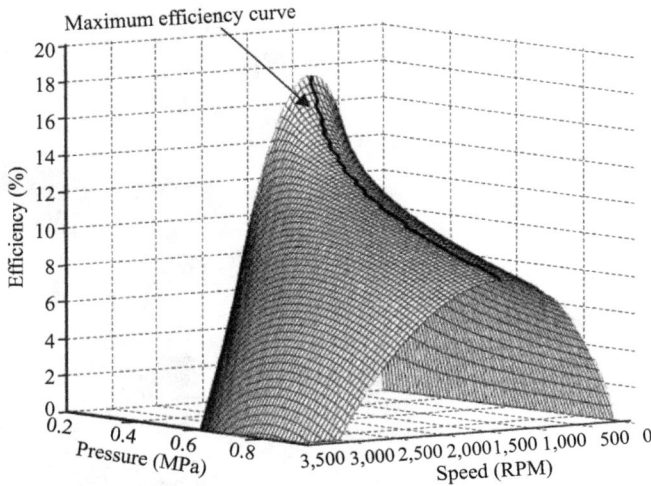

Figure 6.8   Pneumatic motor maximum efficiency line [8]

of several quantities (pressure, flow, speed, current, etc.) in order to determine the ideal speed according to the maximum efficiency line depicted in Figure 6.8 and, thus, use it in the speed control module [8].

Figure 6.8 shows that the pneumatic motor has low efficiency, less than 20%. The reduced efficiency of a small-scale system with pneumatic motor is associated with the heat losses due to the high area/volume ratio of such machines, caused by the irreversibility of a nonadiabatic process. For similar reason and because they have critical internal tolerances to reduce leakage losses, turbines are not used for small-scale power generation.

6.3.2   Hybrid SS-CAES systems

SS-CAES systems are mechanical and hence present slow reaction to transient electrical load changes and have narrow maximum efficiency or power peaks [10]. In sort, a small deviation from the MPPT or MEPT may cause significant loss of efficiency or generated power. Hybrid combinations of an SS-CAES system with other storage devices have the potential to resolve this issue. Supercapacitors or batteries, for example, can be used to buffer fast load fluctuations and enable the operation of SS-CAES systems at their maximum power or efficiency points. Moreover, despite the efforts concentrated on the MPPT and MEPT strategies of SS-CAES systems, the efficiency of pneumatic air motors remains very low [8,9]. Therefore, other conversion solutions have also been evaluated in hybrid SS-CAES systems, as the implementation of oil-hydraulic machines [8]. Some of these systems are discussed in the following.

6.3.2.1 CASCES system

The CASCES system is represented in Figure 6.9. This system consists of the combination of a high-capacity energy storage technology (compressed air) and an auxiliary energy storage device with high-power density (supercapacitor) [8]. The components of the pneumatic/electric assembly are reversible; therefore, the energy stored in this system comes from an external primary source, which may be the power grid or a photovoltaic generator, for example.

The combination of supercapacitors with the compressed air system allows a better quality of the energy delivered to the load, smoothing the imminent fluctuations in the output voltage arising from possible load variations. As can be seen in Figure 6.10, the supercapacitor is employed to supply the load when the motor is at rest and it is also used as an assistant in maintaining the power delivered at peak load times [8].

Figure 6.10(a) illustrates the power curve from compressed air, converted and delivered to the system. The shaded regions correspond to the instants when the pneumatic/electric conversion assembly is supplying the load and charging the supercapacitors. In this working mode, the voltage of the supercapacitor bank is monitored and the pneumatic/electric group is activated when such voltage reaches the minimum value established for the system. It is remained on until the maximum voltage value is reached or until the compressed air resource is exhausted.

In Figure 6.10(b), the shaded parts correspond to the periods in which the supercapacitors supply energy to the load. Moreover, in Figure 6.10(c) there are

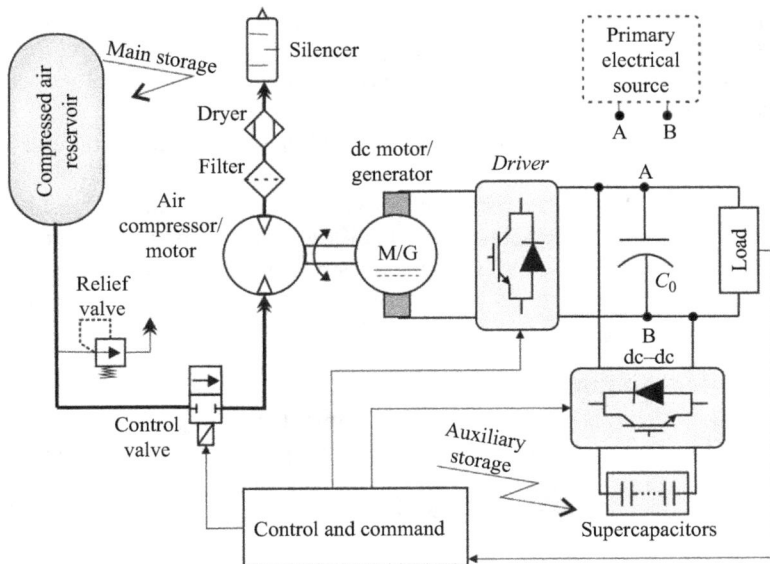

Figure 6.9   CASCES hybrid storage system [8]

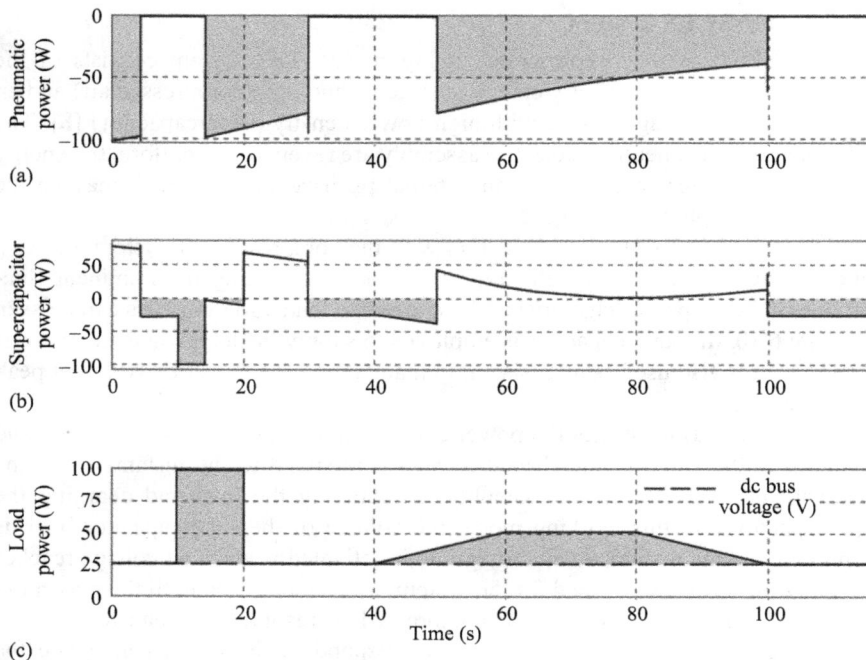

Figure 6.10   Power curves of the CASCES system [8]

two regions of power variation, and at the beginning of these variations, the capacitor transfers all the stored energy to the load and, subsequently, the pneumatic/electric assembly is activated to ensure that the supercapacitors recharge process occurs.

As suggested in Figure 6.10, the support of supercapacitors to the pneumatic storage system can be quite interesting, especially in terms of load power variations. A similar strategy was proposed in [11], but using a battery instead of a supercapacitor bank. Nonetheless, the small-scale pneumatic/electric conversion system remains with the same efficiency restrictions of the pneumatic vane air motor. To increase the overall efficiency of the SS-CAES system, hydropneumatic combinations are suggested to replace the purely pneumatic part [8].

6.3.2.2   BOP-A system

The BOP-A system is a form of energy storage that combines the usage of hydraulic oil and pneumatic devices. The gas compression and expansion processes occur in a closed cycle, i.e., with sealed gas and no air intake/exhaust from/to outside [8,22]. The proposal of this system is based on the high efficiency of the engines and oil pumps. These hydraulic machines operate at high-pressure levels, in the range of 10–35 MPa, and can exhibit efficiencies above 90% at these pressure levels [22].

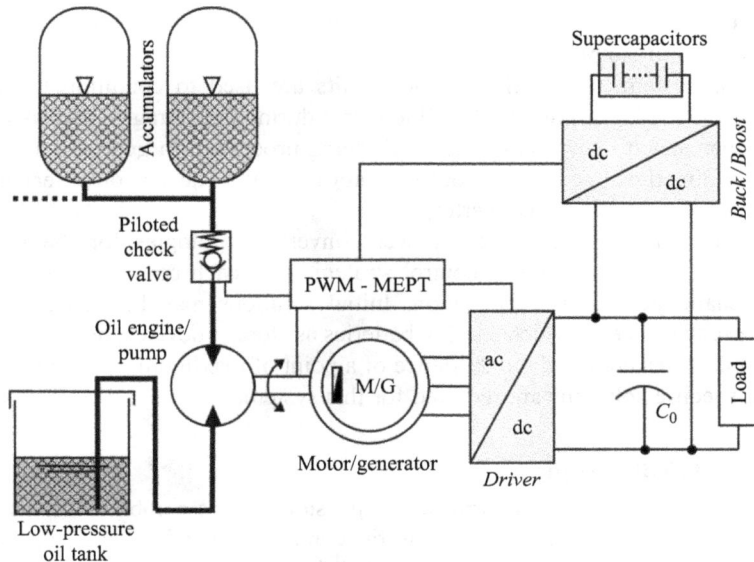

Figure 6.11   BOP-A hybrid system with supercapacitors [8]

The energy storage of the BOP-A system uses pressure reservoirs industrially known as hydraulic accumulators. These accumulators have two chambers, one for the gas and another for the liquid, separated by a membrane or a floating piston. When pumping oil into the accumulator, the pressure in the oil compartment starts to rise and, thus, the membrane/piston moves in order to reduce the volume of the gas chamber, compressing it.

In this type of application, an inert gas such as nitrogen is usually used, since the combination of oxygen and oil under pressure can give rise to an explosive mixture. Figure 6.11 illustrates the accumulators and other basic components of the BOP-A storage system, including a bank of supercapacitors employed as an auxiliary reservoir for the same purpose described in the CASCES systems.

The devices of the BOP-A system are reversible, so when the rotating machine (motor/generator) operates as a motor, it drives the hydraulic machine, which acts as a pump. The hydraulic pump is responsible for transferring oil from a low-pressure tank into the accumulators, compressing the pre-charged gas in the chambers. At the end of this stage, the control valve is closed and the gas stays compressed, storing energy. When the stored energy is requested, the control valve is opened and the compressed gas pushes oil out of the accumulators. The oil flows into the low-pressure tank by moving the hydraulic machine and making it work as a motor that, in turn, moves the generator, converting mechanical energy into electrical energy.

Despite the high efficiency of the hydraulic motors, the overall efficiency of the system is affected by the energy dissipation and the heat exchange resulted from the turbulent fluid flow, the several transformation stages for a complete storage cycle, and power supply.

In this system, power electronics circuits are used to condition the stored energy to the load needs and to drive the motor during the storage stage, as well as in the supercapacitors charging and discharging processes. In general, the system requires bidirectional converters for the motor/generator (dc–ac converter) and for the supercapacitors (dc–dc converter).

The development of suitable power converter topologies for the required power levels and the design of control strategies for each converter can be considered challenges in this application. Initial solutions may be based on those already used in systems which employ batteries as storage devices; however, due to its dynamical behavior and the existence of a point of maximum power that can be tracked, specific solutions are required for this system.

6.3.2.3   BOP-B system

The BOP-B system is also a form of energy storage that combines the usage of hydraulic oil and pneumatic devices, but the compression and expansion processes occur in open cycle; that is, the compressed air is admitted from the ambient during the compression stage and released to the atmosphere at the expansion [8,22,23]. The air compression and expansion processes performed in this system are based on a technology known as liquid piston [24]. Basically, it is a direct hydraulic/pneumatic transformation arrangement which uses a column of liquid in the compression process to squeeze the gas present in a given fixed volume chamber. The liquid fluid, typically water or oil, is driven by a hydraulic pump and pneumatic valves control the air inlet and outlet of the chamber [24,25].

Figure 6.12 depicts the schematic of a BOP-B system with near-isothermal compression/expansion with the support of supercapacitors. During the expansion process, a specific compressed air volume flows into the chamber (1R) of the right cylinder through the opening valve (D). The hydraulic valve (4) is switched to position (a) and thus the fluid from the liquid piston (2R) flows through the heat exchanger (3R), passing through the hydraulic motor and the hydraulic valve (4) in position (a). Thereby, the liquid piston (2L) fills the cylinder on the left, discharging the air from the chamber (1L) to the atmosphere through the silencer (7) due to the opening of the valve (B). After the stroke of the liquid piston (2L) ends, the valve (4) inverts the flow direction by switching to position (b), without changing the rotational direction of the hydraulic motor (5). For this purpose, the valves (B) and (D) have to be closed and the valves (A) and (C) have to be opened. The effects of torque changes during operating cycles can be mitigated by auxiliary storage with supercapacitors [8,22,23].

The BOP-B system also works as a compressor when the fluid is moved through the hydraulic machine, which is operating as a pump. Such hydraulic engine is driven by the electric machine, which runs as a motor when fed by the power grid. Thereby, with the correct valve switching sequence, the air in chambers 1D and 1E can be compressed and sent to the reservoirs.

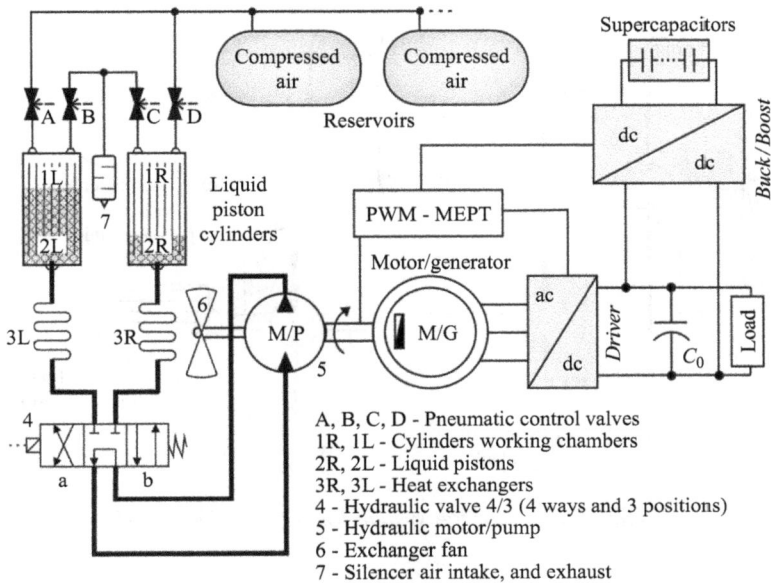

Figure 6.12   BOP-B hybrid system with supercapacitors [8]

In terms of power electronics, this solution differs from the previous one by the dynamic behaviors of the storage system and the MPPT. Therefore, the choice of the power converters and the control strategy should consider these new characteristics of the BOP-B system. Moreover, in [25] and [26], it is possible to observe the extension of its field of application next to renewable sources, where an open-cycle hydropneumatic energy storage system, similar to the described BOP-B system, is proposed for the application in wind turbines, considering liquid piston compression/expansion chambers, as shown in Figure 6.13.

6.3.2.4   BOP systems comparison

Table 6.2 presents an estimated cost comparison among BOP-A, BOP-B, and lead-acid battery storage systems to be used in an application with a photovoltaic generator that supplies a house [22]. The case study has considered an average power generation of 4 kW over a period of 7 h (during the day), i.e., 28 kWh. For the discharge of the system, an interval of 5 h (at night) was considered, with an average power consumption of 4 kW, i.e., 20 kWh. The economic framework also considered a requirement for autonomy of 3 days; therefore, the compared storage systems have been sized to supply 60 kWh.

Table 6.2 shows that the highest storage cost is from the BOP-A system. This is mainly due to its low energy density, of the order of 2.5 Wh/kg. In general, the estimated costs of BOP-type applications are higher than lead-acid batteries, but since a much longer life span has been estimated for BOP systems, the cost of energy (€/kWh) ends up being lower than battery systems [22].

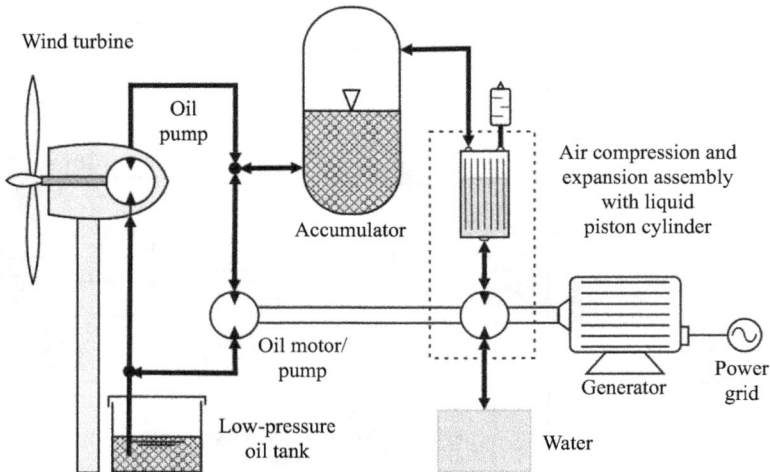

Figure 6.13   Storage system for wind turbine [25]

Table 6.2   Cost evaluation for 60 kWh–10 kW peak storage [22]

60 kWh	Lead acid	BOP-A	BOP-B
Storage technology	€18,000	€60,000	€4,500
Conversion technology	–	€15,000	€18,000
Operation and maintenance	€3,500 (30%)	€15,000 (20%)	€4,500 (20%)
Total cost	€23,500	€90,000	€27,000
Life cycle	210,000 kWh	900,000 kWh	900,000 kWh
Cost of energy	€0.11/kWh	€0.10/kWh	€0.03/kWh

Source: adapted from [22].

The studies presented in [27] also suggest a favorable opinion on small-scale air storage systems combined with the usage of photovoltaic generators, with a payback period of 5–7 years for residential and hotel applications, pointing out that about 40% of the cost is respect to the value of the reservoirs.

6.3.3   Future trends

Based on the literature review, it is possible to understand the applicability of small-scale compressed air storage systems, especially in conjunction with renewable energy sources. Therefore, more research in this field should be stimulated, such as the study of the hybrid configuration of Figure 6.14, based on the BOP-B system, with open-cycle hydropneumatic architecture and supercapacitors, connected to the power grid and powered by photovoltaic generators.

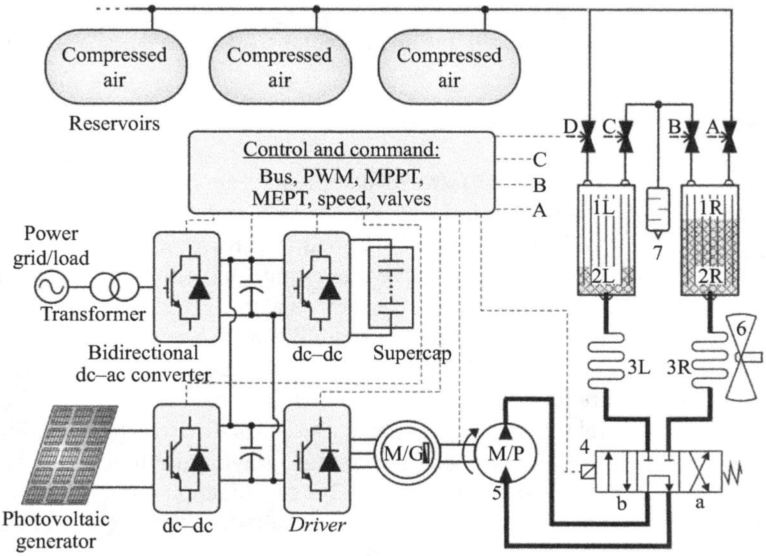

Figure 6.14   Open-cycle air storage system with hydraulic interface and supercapacitors, adapted from [8]

One of the main advantages expected for the system of Figure 6.14 is to take advantage of the better performance of the hydraulic motor with a higher energy density when compared to the configuration of BOP-A system. It is worth noting that the BOP-A system needs a tank to store oil at low-pressure, and its volume is directly related to the energy storage tanks, since about 50% of the hydraulic accumulator's volume is occupied by oil in the compression stage.

The arrangement of Figure 6.14 requires a mechanical interface for linking oil and air fluids, whose operating cycle is identical to the BOP-B system, previously described. The electromechanical transformations are reversible, i.e., performed by a rotating machine that can operate as motor or generator. This bidirectionality is also required for the motor/generator driver, the dc–dc converter dedicated to the bank of supercapacitors, and the bidirectional dc–ac converter between the dc bus and the power grid. In the inverter mode, the converter enables the supply of electricity from the photovoltaic array and recovered it from the storage system (supercapacitors and pneumatic/hydraulic interface) to the grid. In the rectifier mode, the converter can be used to recharge the storage system, at the time of low demand or lack of photovoltaic generation (night periods).

The definition of the converter structures will depend on the power to be processed. The control strategy should optimize the use of the stored energy (using MPPT or MEPT techniques), consider the system dynamics, and control all operating modes (power flow direction), such as photovoltaic generation, charging and discharging of supercapacitors, charging and discharging of the compressed air system, and power injection into the grid. In addition, studies on the overall efficiency of the system are required, considering restrictions such as operation cost, efficiency, life span and uninterrupted operation.

6.4   Final considerations

Energy storage strategies in the form of compressed air have advanced for both large- and small-scale systems. In large-scale CAES systems, advances in energy density levels and efficiency gains of up to 30% are presented, thanks to the mechanisms and intermediate stages of heat exchange and reuse of A-CAES and I-CAES systems. The evolution of these systems is also well seen in environmental aspects due to the minimization of fuel burning during the process.

Studies of SS-CAES systems reveal that pneumatic microturbines typically operate at low-pressure levels and have low efficiency, features that directly affect the efficiency of purely pneumatic systems. As an alternative to overcome this issue, hybrid systems (oil/air) are approached using hydraulic motors/pumps with better efficiency and higher-pressure levels. This combination requires an interface between fluids, oil and air, which makes the system more complex than the purely pneumatic ones.

Power electronics are in the various application possibilities of small-scale systems, acting as a link between the storage system and the power grid, in the presence of isolated loads and renewable energy sources. Bidirectional converters, modeling, and multilevel control strategies will be needed to the development of SS-CAES. The improvement of these systems requires and stimulates the development of the various elements that compose it, as a way to optimize each one of them and increase the overall efficiency.

References

[1]   Vazquez S., Lukic S.M., Galvan E., Franquelo L.G., and Carrasco J.M. "Energy storage systems for transport and grid applications." *IEEE Transactions on Industrial Electronics.* 2010;57(12):3881–3895.

[2]   Mohd A., Ortjohann E., Schmelter A., Hamsic N., and Morton D. "Challenges in integrating distributed energy storage systems into future smart grid." *Proceedings of the IEEE ISIE*; Cambridge, UK, 2008. pp. 1627–1632.

[3]   Xiaoguang C., Chenghui Z., Li K., and Yefei J. "Dynamic modeling and efficiency analysis of the scroll expander generator system for compressed air energy storage." *Proceedings of the IEEE ICEMS*, Beijing, China, 2011. pp. 1–5.

[4]   Perkins D.E. "Compressed air energy storage (CAES)," in Capehart L.B. (ed.). *Encyclopedia of Energy Engineering and Technology,* London: Taylor & Francis Group; 2017, vol. 3, pp. 214–218.

[5]   Cleary B., Duffy A., O'Connor A., Conlon M., and Fthenakis V. "Assessing the economic benefits of compressed air energy storage for mitigating wind

curtailment." *IEEE Transactions on Sustainable Energy.* 2015;6(3): 1021–1028.

[6] Khamis A., Badarudin Z.M., Ahmad A., Rahman A.A., and Hairi M.H. "Overview of mini scale compressed air energy storage system." *Proceedings of the PEOCO*; Shah Alam, Malaysia, 2010. pp. 458–462.

[7] Rogers A., Henderson A., Wang X., and Negnevitsky M. "Compressed air energy storage: Thermodynamic and economic review." *Proceedings of the IEEE PES General Meeting*; National Harbor, MD, USA, 2014. pp. 1–5.

[8] Lemofouet S. and Rufer A. "A hybrid energy storage system based on compressed air and supercapacitors with maximum efficiency point tracking (MEPT)." *IEEE Transactions on Industrial Electronics.* 2006;53(4): 1105–1115.

[9] Kokaew V., Sharkh S.M., and Torbati M.M. "Maximum power point tracking of a small-scale compressed air energy storage system." *IEEE Transactions on Industrial Electronics.* 2015;63(2):985–994

[10] Kokaew V., Torbati M.M., and Sharkh S.M. "Maximum efficiency or power tracking of stand-alone small-scale compressed air energy storage system." *Energy Procedia.* 2013;42:387–396.

[11] Omsin P., Sharkh S.M., and Torbati M.M., "A hybrid SS-CAES system with a battery." *Proceedings of the ECTI-CON*; Pattaya, Thailand, 2019. pp. 159–162.

[12] Atlas Copco Airpower NV, *Compressed Air Manual.* 8th edn, 2015.

[13] Saravanamuttoo H.I.H., Rogers G.F.C., Cohen H., Straznicky P.V., and Nix A.C. *Gas Turbine Theory.* 7th edn. Harlow: Pearson; 2017.

[14] Budt M., Wolf D., Span R., and Yan J. "A review on compressed air energy storage: Basic principles, past milestones and recent developments." *Applied Energy.* 2016;170:250–268.

[15] Kim Y.-M., Lee J.-H., Kim S.-J., and Favrat D. "Potential and evolution of compressed air energy storage: Energy and exergy analyses." *Entropy.* 2012; 14:1501–1521.

[16] Bullough C., Gatzen C., Jakiel C., Koller M., Nowi A., and Zunft S. "Advanced adiabatic compressed air energy storage for the integration of wind energy." *Proceedings of the European Wind Energy Conf. (EWEC)*; London, UK, 2004. pp. 1–8.

[17] RWE Power AG. *ADELE—Adiabatic compressed-air energy storage for electricity supply* [online]. 2010. Available from: http://www.rwe.com/web/cms/mediablob/en/391748/data/235554/1/rwe-power-ag/press/company/Broch ure-ADELE.pdf [Accessed Aug 3, 2020].

[18] Geissbuhler L., Becattini V., Zanganeh S., *et al.* "Pilot-scale demonstration of advanced adiabatic compressed air energy storage, Part 1: Plant description and tests with sensible thermal-energy storage." *Journal of Energy Storage.* 2018;17:129–139.

[19] Wang J., Lu K., Ma L., *et al.* "Overview of compressed air energy storage and technology development." *Energies.* 2017;10:991.

[20]   Proczka J.J., Muralidharan K., Villela D., Simmons J.-H., and Frantziskonis G. "Guidelines for the pressure and efficient sizing of pressure vessels for compressed air energy storage." *Energy Conversion and Management.* 2013; 65:597–605.

[21]   Villela D., Kasinathan V.V., Valle S., *et al.* "Compressed-air energy storage systems for stand-alone off-grid photovoltaic modules." *Proceedings of the IEEE PVSC*; Honolulu, HI, USA, 2010. pp. 962–967.

[22]   Lemofouet S. and Rufer A. "Hybrid energy storage system based on compressed air and super capacitors with maximum efficiency point tracking." *Proceedings of the IEEE EPE*; Dresden, Germany, 2005. pp. 1–10.

[23]   Cyphelly I., Rufer A., Briuckmann P., Menhardt W., and Reller A. *Usage of Compressed Air Storage Systems.* Swiss Federal Office of Energy, 2004.

[24]   Van de Ven J.D. and Li P.Y. "Liquid piston gas compression." *Applied Energy.* 2009;86(10):2183–2191.

[25]   Saadat M., Li P.Y., and Simon T.W. "Optimal trajectories for a liquid piston compressor/expander in a compressed air energy storage system with consideration of heat transfer and friction." *Proceedings of the American Control Conf. (ACC)*; Montreal, QC, Canada, 2012. pp. 1800–1805.

[26]   Saadat M., Shirazi F.A., and Li P.Y. "Modeling and control of an open accumulator compressed air energy storage (CAES) system for wind turbines." *Applied Energy.* 2015;137:603–616.

[27]   Tallini A., Vallati A., and Cedola L. "Applications of micro-CAES systems: Energy and economic analysis." *Energy Procedia.* 2015;82:797–804.

Chapter 7

Compressed air energy storage systems, towards a zero emissions in electricity generation

Bernardo Llamas[1], Eva M. Blanco-Brox[1], María C. Castañeda[1] and Gabriel Barthelemy[1]

In the fight against climate change, the electricity sector is involved in the promotion of renewable sources. These technologies, free of CO_2 emissions in their electricity generation process, suffer from a low load factor. This requires expensive backup systems for the power grid to guarantee supply. To increase the availability of these resources, the concept of 'massive energy storage' arises, considering chemical storage in the form of hydrogen, pumping water, or storage of compressed air in the subsurface.

This chapter will review the concepts of this latest technology, gathering the concept of compressed air storage taking advantage of obsolete infrastructure and a novel alternative to thermal energy management. The viability of this technology depends on an adequate use of the thermal energy exchanged in the compression and expansion processes, which allows an increase in the overall efficiency of the process and an improvement in the competitiveness, compared to other mass storage solutions.

Keywords: Energy storage; renewable energy; compressed air energy storage; hybrid storage systems; BIO-CAES technologies; adiabatic CAES

7.1 Introduction

The current economic model, based on fossil fuels [1], increased CO_2 concentration in the atmosphere since the industrial era (Figure 7.1(a)), and a high risk of modifying the climate on the planet, with consequences not yet predictable. The concentration of CO_2 in the atmosphere continues to increase faster each year [2]. The successive economic (2009) and health (2020) crises have not slowed the rise of emissions. Indeed, CO_2 concentration in the atmosphere will be the highest of the last 10 years in 2020 (Figure 7.1(b)).

[1]Universidad Politécnica de Madrid, ETSI Minas y Energía, Madrid, Spain

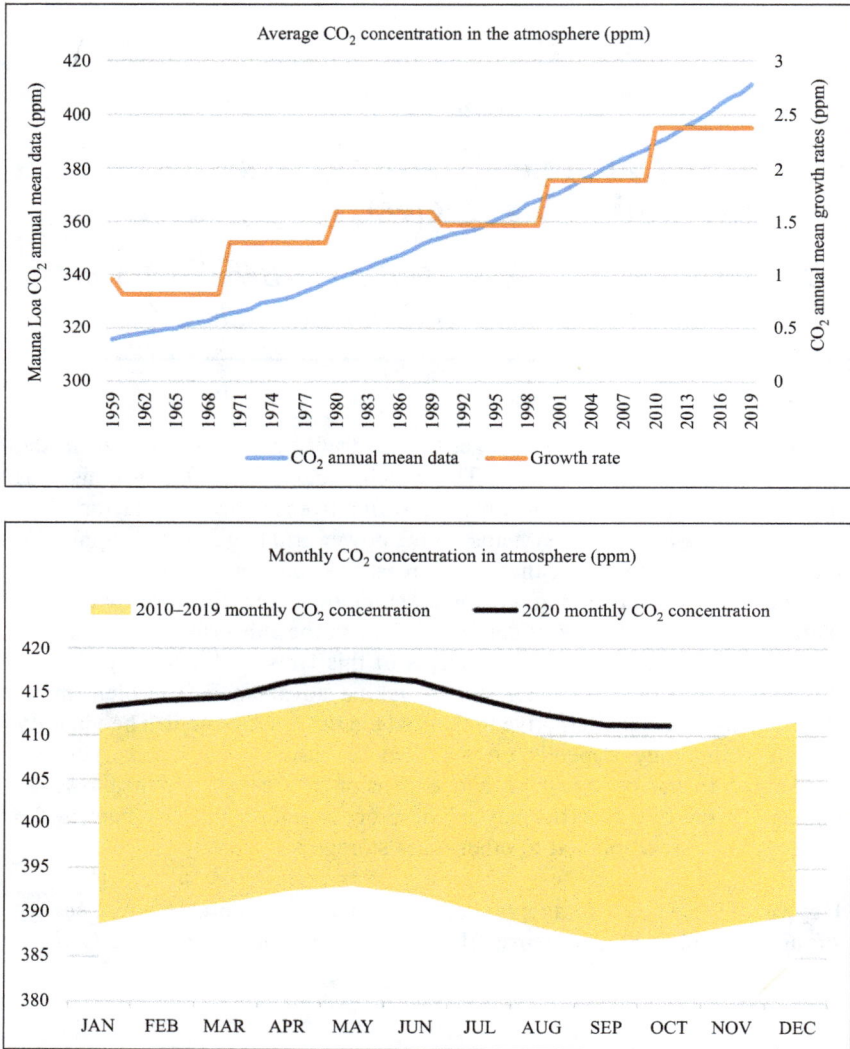

Figure 7.1   (a) CO$_2$ concentration in the atmosphere and annual mean growth rate. (b) Monthly CO$_2$ concentration in atmosphere: 2010–19 period and 2020 year (Mauna Loa Observatory, NOAA, 2021)

Politically, the United Nations promoted different protocols to regulate anthropogenic emissions of greenhouse gases (GHGs), such as the Paris Protocol (COP21) [3]. It established that the limit of global temperature increase should be below 2 °C and it advises not to reach an increase higher than 1.5 °C, compared to the pre-industrial average temperature.

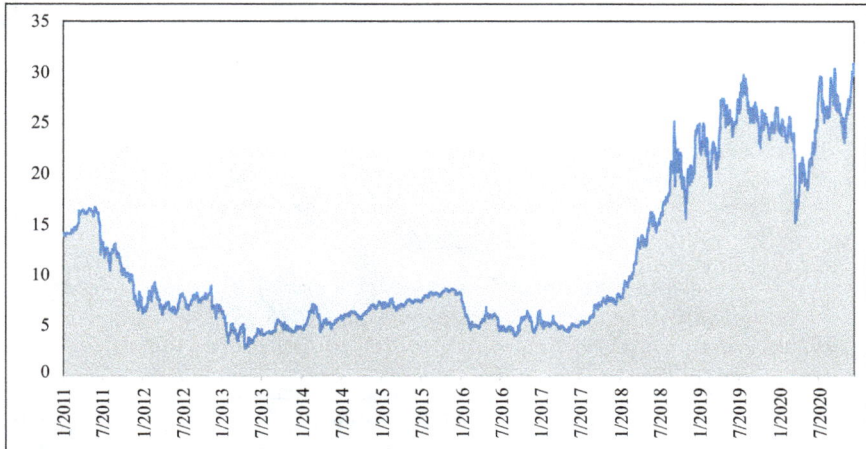

Figure 7.2 CO_2 cost evolution in the EU Emissions Trading System (€/tCO₂)

The European Union (EU) established a directive package to establish an Emission Rights market and, therewith, promote compliance with the Kyoto Protocol (Directive 2003/87/EC). This market regulates different industrial sectors (electricity generation, oil refining, cement manufacturing, ceramics, glass, and paper), to which was later added (2009/29/EC) the need to report CO_2 emissions from the aviation sector.

Indeed, subsequent directives (2018/410/EU) and objectives set by the EU [4] have perfected this CO_2 emissions regulating mechanism, and at the end of the third period, the price of CO_2 is around €25–30/tCO₂ (Figure 7.2).

The next phase of this market (phase 4: 2021–30) foresees an even more significant reduction of emission rights in the market: 2.2% compared to the current rate of emission rights reduction (1.74%). This will continuously maintain and increase the cost of rights over time, accelerating the transition towards a neutral industry in CO_2 emissions.

The electricity generation sector has been one of the main emitting sectors, with technology based on fossil fuels. However, in the last decade, the generation mix has been transformed towards a model based on renewable sources [5]. Wind power and photovoltaic energy have experimented strong growth (Figure 7.3).

Nevertheless, these technologies, without CO_2 emissions, create an additional problem: security of supply. They are sources considered intermittent due to the impossibility to predict the behaviour of primary energy sources (wind, solar radiation). Thus, they must be achieved with backup technologies that guarantee the supply of electrical energy.

Given this situation, there is a need to store energy in a massive way, considering the pumped hydroelectricity storage (PHS), compressed air energy

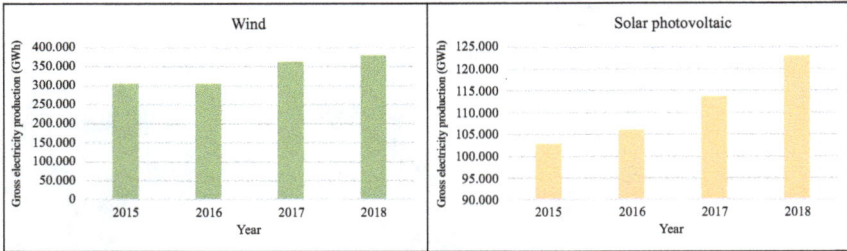

Figure 7.3   Growth in total gross domestic production of EU wind and solar photovoltaic energy between 2015 and 2018 [5]

storage (CAES), and hydrogen (H_2). In fact, these massive energy storage systems in any of their three forms partially solve the relationship between the production of electricity from renewable energy sources and the demand for electricity consumption.

In this chapter, CAES technology will be evaluated, with special emphasis on the current limitations and how alternatives have been proposed to overcome this limitation: (i) high cost of subsurface exploration and characterization and (ii) lower energy efficiency.

7.2   Massive energy storage technologies

Three mass storage technologies will be considered based on high energy storage capacity; using generation power and time of use as criteria, considering the energies that reach, at least, 50 MW of power that can be sustained for hours. As discussed in the introduction, these technologies are hydrogen as an energy storage vector (H2), PHS, and CAES.

Table 7.1 presents the energy efficiency, disclosed as a percentage that can be obtained through the different energy storage systems: electrical, chemical, thermal, and mechanical. Each one has advantages, limitations, and different applications; this will determine which one to use. Within the type of mechanical energy storage, PHS is the most efficient technology currently.

7.2.1   Hydrogen, as an energy storage vector

The interest in the hydrogen economy is not new [9]; it grew in the last decades of the twentieth century [10]. Recently, interest in developing hydrogen as an energy vector free of CO_2 emissions has led to increased funding to research projects. Hydrogen production can be classified according to four main routes: renewable, non-renewable, nuclear, or biomass [11]. Currently 95% of the hydrogen produced has fossil origin [12].

Most hydrogen used today is known as grey, brown, or black, depending on whether it is produced from the reforming of natural gas, coal, or oil. As a result, when producing this, carbon dioxide (CO_2) is released into the atmosphere. When

Table 7.1   *Energy efficiency in different energy storage technologies: average percentage of all types of storage within the four large groups [6–8]*

	Energy efficiency
Electrical energy storage	95%–98%
Chemical energy storage	Ni–Cd: 60%
	Li-ion: 98%
Thermal energy storage	20%–45%
Mechanical energy storage	CAES: 50%; A-CAES: 70%
	PHS: 70%–85%

part of this carbon dioxide is captured and stored, it is known as blue hydrogen, but not all the carbon dioxide released during the process can be captured, and there are problems to store it. There is also turquoise hydrogen, produced from natural gas through the pyrolysis process, but because it is a fossil fuel and is not 100% emissions-free.

However, the goal is to produce hydrogen from renewable sources (green hydrogen), based on water electrolysis [13]. Therewith, produced and non-consumed electricity can be stored in the form of hydrogen. Typically, the process consists of an electrolyser, a storage system, and a fuel cell [14].

Hydrogen can be stored both on the surface (storage tanks) and underground in geological deposits. For the first choice, there are hydrogen storage processes by compression, cryogenic liquid, and metal hydrides. Due to the costs associated to these technologies, the latter is typically used for small amounts of hydrogen [15]. Using the subfloor as a hydrogen storage site allows high capacity (above 5 GWh) to ensure long-term stability in the network. Not only in terms of capacity, underground installations have more advantages over surface installations, but also greater safety, lower costs, better space management, and high availability of suitable geological structures [16].

Although storing hydrogen underground is presented in a similar way as natural gas, the behaviour of this element and the reactions that could occur must be considered [17]. Storing hydrogen as a compressed gas is the simplest form. This is because no transformation or special condition is necessary, and only a compressor system to reach the optimum pressure is needed [18]. Of all the geological structures typically considered for the storage of energy cavities in saline formations, abandoned mines, and porous formations [19], cavities in saline domes are an ideal choice due to the low permeability of the gas and the favourable rheological properties [20].

Finally, fuel cells allow electrochemical reactions to stop transforming hydrogen into energy, as well as heat as a co-product. It is the commonly used technology, where hydrogen is used as fuel and air oxygen as a fuel and the product of this reaction is water [21]. Up to six types of technologies can be identified: proton-exchange membrane fuel cell, direct methanol fuel cell, alkaline fuel cell, phosphoric acid fuel cell, molten-carbonate fuel cell, and solid oxide fuel cell.

7.2.2   Pumped hydroelectric storage

Currently, PHS accounts for approximately 99% of the world's storage capacity with 158 GW in 2019 [6]. Therefore, it remains an important option, particularly for storage of 10–15 h, with a global pump storage capacity that will almost double to 300 GW by 2070 [1]. PHS systems store excess electricity (valley hours) by pumping water from one tank to higher ones. By releasing water (peak hours) through the turbine, stored energy is regained. The efficiency of this energy storage system is high, between 70% and 85% (Table 7.1, [7]), and even 90%, depending on the technology used, the plant size, and the difference in height, in addition to the turbines and the size of the gates [22].

Most conventional facilities are installed in river areas, which means they are defined by water availability, flood control efforts, and increasingly stringent environmental considerations, and therefore have limited potential in future development.

Current research on PHS focuses on the following lines: (a) current state of development of technology by analyzing the advantages and disadvantages of its implementation in developed and developing countries with high energy needs [23]; (b) integration of power converters to transform the conventional (fixed) PHS to one with variable speed, offering flexibility to operate beyond the nominal hydraulic limits [24]; (c) integration into mini-networks from a smaller development approach and operating in conjunction with renewable (intermittent) energies such as wind and solar [25]; (d) methodologies for selecting new, more sustainable sites, a key influence factor on the plant's ability to sustainably provide the expected benefits throughout its life cycle [26].

7.2.3   Compressed air energy storage

CAES technology consists mainly of using non-demanded electricity to store it in the form of high-pressure compressed air. To date, different CAES concepts have been developed to improve the technical and economic performance of the system and adapt to the energy storage needs of different scenarios. To this end, two main problems have been addressed: First, the high risk arising from the exploration and characterization of the subsurface that hinders the development of the reservoir which must provide safety and stability in the air storage and its compression process. Mining cavities, saline domes, and deep aquifers are considered in the subsurface, while surface facilities may be defined as tunnels, or natural gas pipelines. They are classified according to cost ($/KWh gen.), available pre-information, and industry references.

Cavities in saline formations are one of the most interesting ones for their physical properties and their low cost of construction [27]. However, abandoned mines might play a significant role as the storage facility is already constructed, and facility investment is reduced (SMART MinEnergy project). To select the most suitable cavity, it is advisable to establish a detailed selection method and characterization of structures, to minimize the mentioned risk. Multi-criteria algorithms can be used to establish a hierarchy of alternatives and identify structures with the greatest potential [28].

The low energy efficiency of the system is the latest point. The thermal energy generated in the compression stage should be managed to use it in the air expansion stage. Some concepts have been defined such as adiabatic CAES (A-CAES) and isothermal CAES (I-CAES) [29]. Unlike conventional CAES (efficiency of 50%), A-CAES stores thermal energy to heat expansion air during the discharge process. This improvement not only optimizes the level of energy utilization of the system, but also eliminates fossil fuels consumption. Cycle efficiency can reach up to 60%–70% [30]. The I-CAES reduces thermal energy loss by cooling the air during the compression process to prevent temperature increase while using recycled compression heat in the release process to maintain a constant temperature. The expected efficiency is more than 80% [31].

In addition, some new technologies are based on CAES technology, such as liquid air energy storage (LAES). It is based on the principle of storing energy as cryogenic air in liquid state, which can be contained in pressure vessels close to room temperature [32].

However, all these advances are integrations designed on a large scale. Some research activities are currently exploring the application of CAES on a smaller scale [33–36]. This concept has several advantages: lower environmental impact and exploratory risk; lower investment and maintenance cost; and better operating capacity.

For years, several CAES-related projects have been moving forward around the world, especially in the USA (Norton-Ohio, ISEP-Iowa, Seneca-Eastern US, Bethel and Matagorda Energy Center-Texas, Columbia Hills, and Yakima Minerals – North Western US) and Ireland (Larne-Northern). UK, Denmark, Germany, and Japan have also developed projects that are in the early stages of planning [37].

7.2.4 Technology comparison

The objective of mass energy storage as a means of achieving the decarbonization of the system is to obtain a high efficiency and viability of renewable energies: expectation of generation parallel to the demand for consumption. While hydrogen energy storage technology (H_2) is being developed with good success rates, it is not able to exist industrially as it uses small electrolysis for production.

Among the most outstanding technologies for mass energy storage, water pumping (PHS), and CAES (Table 7.2) are the only ones that allow large amounts of energy (above 100 MWh) to be stored in the long term (longer service life). However, the common disadvantage is the high cost of capital involved in the main components.

Between these two options, PHS is technically more developed, and has higher capacity and lower operating and maintenance costs than CAES. However, energy storage in the form of compressed air avoids the main PHS limitations: geographical site restriction (in mountainous regions), the high environmental impact, and the use of water. In addition, CAES technology procures a long period of

Table 7.2   *Comparative technology analysis for PHS, CAES, and H$_2$*

		PHS [21]	CAES [38]	Hydrogen + fuel cell [39]
Power capacity	MW	100–5,000	100–1,000	50
Discharge time at rated power		Hours–days	Hours–days	Minutes–hours
Life time	Years	40–60	20–40	5–20
Energy efficiency	%	70–85	40–70	20–50
Power capital cost	$/kW	500–4,000	400–1,000	500–10,000
Energy capital cost	$/kWh	5–430	2–30	2–15

storage, and entails affordable costs and high efficiency (especially in new improvement developments). The power generated through this storage system varies between 100 and 1,000 MW and has a much higher capacity compared to other systems, except for PHS. Either way, it seems reasonable to consider that both solutions can coexist given the geographical conditions, which may imply the selection of one or the other technology based on hydraulic availability, the possibility of potential compressed air storage, or the existence of renewable electricity generation facilities.

In short, combination of CAES and renewable energy is considered one of the most promising solutions to address renewable energy issues. Its sustainability index shows a positive value in comparison with other technologies [40]. In addition, due to its excellent capacity and extensive operating conditions, CAES can also provide ancillary services such as frequency control, reserve, and reactive power regulation for network systems. CAES is considered one of the most effective means for network-attached storage and can be cost-effective with load change applications based on the difference in electricity price at valley peaks [39,41].

7.3   Compressed air energy storage

The different CAES technologies, based on the use of surplus electrical energy – ideally from renewable energies – through air compression processes and its underground storage, will be considered. It is selected considering the thermodynamic aspects of air energy in the compression and expansion processes. It is diabatic CAES (D-CAES) if it does not exploit the air and does not cede air to the atmosphere; it is A-CAES or I-CAES if it is exploited, depending on the type of process. Finally, LAES, which allows the quantity and energy volume to be increased, will also be studied.

7.3.1   *Diabatic CAES*

The CAES system has two parts: compression (compression train) and discharge (turbine). During times of lower electrical demand, air pressure storage is done by

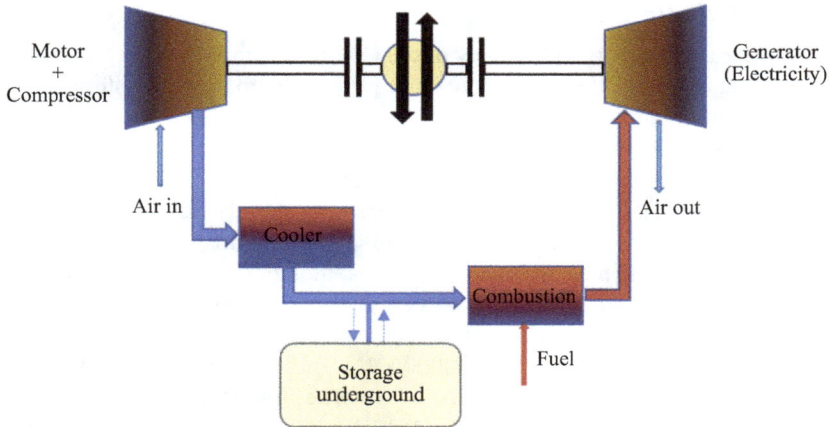

Figure 7.4 Scheme of operation of a CAES under operating conditions to accompany generation and demand

using compressors, which convert electrical energy into potential. With better precision, it can convert the exergy into pressurized air. When the grid requires electricity to meet increased demand, the air is removed from the cavity, heated in a combustion chamber, and directed towards the expansion train so that the turbines together with the generator produce electrical power (Figure 7.4).

In the compression process, a large amount of heat is generated, which is lost directly into the atmosphere by cooling the air in the compression system coolers. As a result, in the expansion phase, air heating in case of a conventional CAES is diabetic and air must be reheated using fossil fuels. This heat loss makes the efficiency of the cycle around 50% [42,43].

There are two commercial CAES facilities: the first was installed in Huntorf in 1978, with a capacity of 290 MW, to provide service and meet the maximum demand, while keeping the capacity factor of a nuclear power plant constant. The second plant (McIntosh, Alabama) started its operation in 1991. This plant has a capacity of 110 MW, with an energy storage capacity of 2,700 MWh, and it can continuously provide a maximum power for 26 h [44]. Both plants use natural gas to heat the air before the expansion stage. This technology cannot be considered zero GHG emissions. For the first plant, to generate 1 kWh of electricity, it needs 1.6 kWh of natural gas. McIntosh plant has a higher energy efficiency (54%): 0.69 kWh electricity needs 1.17 kWh of natural gas [45]. Additionally, Apex is aiming to provide 317 MW conventional CAES system in Texas [46].

The incentive to develop CAES technology is to achieve energy sustainability and reduce emissions, so the development of conventional CAES aims to avoid the use of fossil fuels and increase the energy efficiency.

7.3.2 Adiabatic CAES

First, to increase the energy efficiency of the CAES system [47], and to avoid fossil energy consumption at the expansion stage [48], a thermal energy storage (TES) system generated at the air compression stage is proposed.

Therefore, to do this, it is proposed to store the heat generated by the TES process, and then recover it and increase the air temperature before the expansion process (Figure 7.5). There are different types of materials that store heat, which can be solid materials, such as sand, concrete, or ceramic materials. They can also be liquid materials, such as organic compounds, molten salts, or alloys [49].

The storage of adiabatic compressed air energy has received a lot of attention in recent years due to its merits of non-consumption of fossil fuels, low costs estimated between €1,200/kW and €2,000/kW, rapid start, wide-range load capacity, and efficiencies that can exceed 70% [50]. The first plant proposed to be built with this technology is the so-called 'ADELE' and 'ADELE-ING' under the projects in Europe (Saxony-Anhalt in Germany; 2010–present). The plant will have a storage capacity of 360 MWh and an electrical power of around 90 MW [37,51]. However, first operative A-CAES plant is located in Ontario (Goderich A-CAES Facility) with a stored capacity of more than 10 MWh.

Some activities focused on improving process efficiency can be described in the following:

• Study of an integrated A-CAES plant with a TES system based on a compact bed of solid material, proposing a new configuration and mode of operation [37].
• Installation of A-CAES with a combined cycle gas turbine plant [52].
• Development of an A-CAES reserve capacity model considering its working mode conversion process, dynamic characteristics, air pressure limits, and thermal storage capacity limits, to optimize the system with reserve scheduling [53].

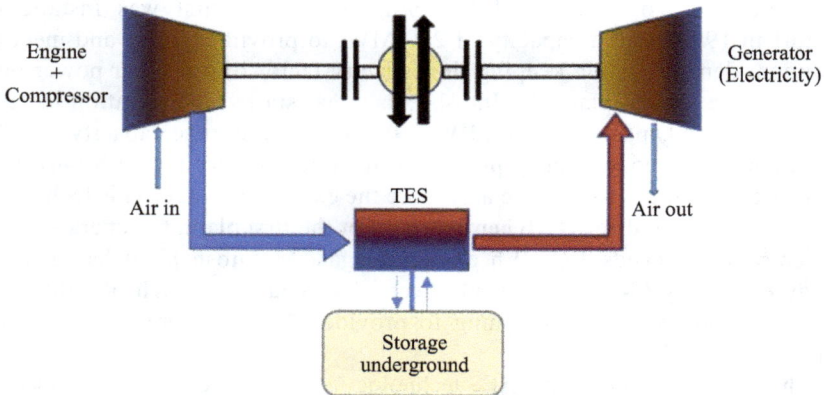

Figure 7.5   Outline of the most important elements of an A-CAES plant

7.3.2.1 Thermal energy storage

TES systems can store heat or cold for use in subsequent industrial processes [54], as would be in nuclear and solar concentration or geothermal plants. TES systems are divided into three types: sensitive heat, latent heat, and thermochemical [55]. Of the three options, the first of these are mature technologies, while the latter is still in a theoretical state. TES technology is more complex than electricity storage technology, since the energy quality is lower, in addition to its subsequent use being more difficult and, therefore, implies a greater loss of energy [43].

Therefore, water is typically used as a fluid, as well as molten salts and other media, on an industrial scale. TES applications exist in the energy and industrial facilities, among others: for example, it is possible to describe TES and a solar energy system for drying solutions [56,57].

Alternatively, it is possible to consider seasonal underground thermal storage [58]. In this case, storage is considered using aquifer TES (ATES) or borehole TES (BTES). In the first case (ATES), water from the target aquifer is extracted and heated before re-injection; in this case, the thermal energy is stored in the water of the aquifer (groundwater). In case of BTES, the probes reach depths of 60–200 m. There are several references to this technology, the largest being the one developed at the Lulea University of Technology [59]. For favourable geological conditions, BTES technology is advantageous for long-term storage from a technical and economic point of view. However, serious environmental aspects should be considered to avoid any impact on groundwater or surrounding buildings.

Considering CAES technologies, TES technology is used as a subsystem to improve the energy efficiency of the assembly [8]. During the compression process, typically in multiphase, temperatures of up to 685 °C can be reached for compression pressures of 50 bar [60]. Considering the storage of thermal energy by storing sensitive heat, two designs can be considered [61]: (a) passive system, using water or oil as a warehouse fluid [62]; (b) active system, using a bed of rock or ceramics [63,64]. Also, it is feasible to consider the method of storage by latent heat (phase change material, PCM) [65].

7.3.3 Isothermal CAES

In case of I-CAES (Figure 7.6), the heat generated in the compression process is continuously extracted, and conversely, the heat is added (in the same way) in the

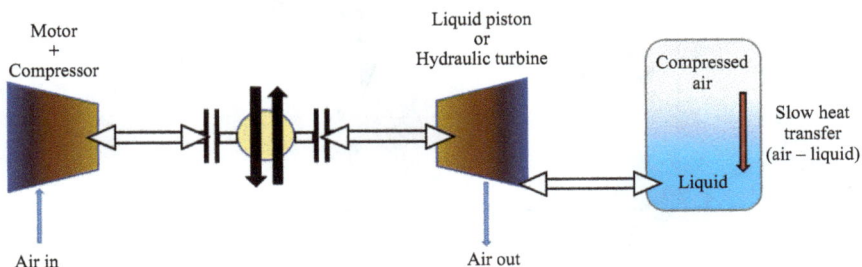

Figure 7.6 I-CAES system

expansion process. This solution allows temperatures throughout the process to remain ideally constant by reducing energy consumption in the compression process and avoiding the need to provide external heat (fuel) prior to expansion [66]. Heat of compression is captured through drops of water or a mixture of air and water. Therefore, the energy is stored in the form of cold air at high pressure and as hot fluid at low pressure.

Therefore, to maximize heat transfer during the compression or expansion process in the isothermal method, liquid pistons or hydraulic turbines are used to carry out these operations. The important thing about this system is that compression and expansion processes happen slowly enough for the heat to transfer between air and liquid [67].

In theory, in I-CAES technology, compression power consumption is equal to the expansion work, so the ideal cycle efficiency is 100%. This process is comparable to the Ericsson cycle [68], in which there are two distinct external combustion processes and four phases: (1) isothermal compression, (2) isobaric heating, (3) isothermal expansion, and (4) isobaric cooling achieving maximum performance (Figure 7.7). This is due to its reversibility property. Therefore, it's basis of the fluid evolution process, the liquid in case of I-CAES, which performs two isothermal and two other isobaric transformations [29].

No I-CAES project has been carried out yet due to the following: (i) the devices needed to perform isothermal heat exchange have not been developed in commercial layout and are still in the research stage; (ii) compared to other CAES, the volumetric density of energy storage is lower and therefore increases the cost per unit of energy storage. The most recent research focuses on resolving the second of the listed handicaps, proposing an open-type I-CAES system in which the time ratio in compression is increased and consequently increases the volumetric density of energy storage [69].

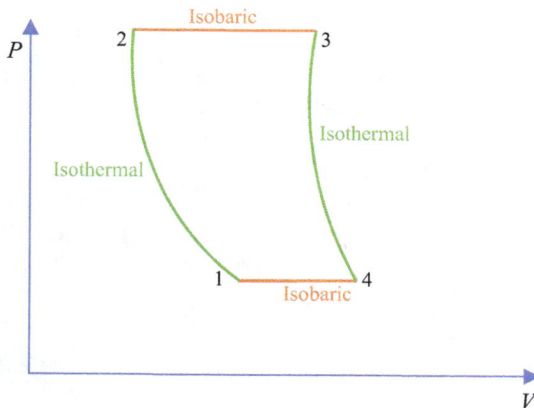

Figure 7.7   Ericsson cycle

7.3.4 Liquid air energy storage

Likewise, technological efforts have led to the development of new CAES-based concepts such as LAES, which is a type of CAES system that aims to increase the amount and volume of energy storage [70], with a higher energy density. The system is based on thermoelectric processes. Energy is stored as a temperature difference between two thermal storages [71]. Therefore, energy storage is carried out by cryogenic air, in a liquid state, through pressure vessels at a temperature close to the environment. This form of storage solves one of the main handicaps of the CAES system: the exploratory risk associated with subsurface exploration for cavity design and construction [7,28].

The storage process can be divided into three phases: (1) liquefying air, which is stored at low pressure in isolated tanks (cryogenic state) – the air liquefaction process is based on the Claude cycle to the detriment of the Linde cycle, for its greater efficiency; (2) when energy is needed, the fluid is compressed out of the storage tank; and (3) heated (environmental thermal energy) and expanded through a turbine (Rankine cycle). If a TES system (cold and heat) is incorporated, the process increases its efficiency [72,73].

Different modifications have been proposed [72,74], as well as its application to a power generation scenario based on intermittent renewable sources [75] or its integration into hybrid energy systems [70,76].

7.4 Analysis of alternatives on hybridization of technologies

7.4.1 CAES coupled with hydrogen system

Bartela [77] proposes an innovative energy storage system that combines the ideas of CAES and hydrogen energy storage (with its subsequent conversion to synthetic natural gas). The efficiency of this system can reach up to 40%: for the assumed operating parameters, the proposed storage energy achieved an efficiency value of 38.15%, which means that the technology is competitive with pure hydrogen energy storage technologies.

Thermal integration of two subsystems allows for efficient storage of large amounts of energy based on the use of pressure tanks with limited volumes. The Bartela model [77] considers three integrated subsystems consisting of a hydrogen generator, methanation, and oxy-combustion processes.

This system avoids the need for fossil fuels (as opposed to D-CAES) and thermal energy storage (A-CAES). However, its energy efficiency is lower than other storage systems considered, such as electrochemical methods (around 90%), pump storage plants (efficiencies of 70%), or the A-CAES systems (these systems efficiencies can reach 70%) [78,79].

The development of the CAES system idea is the result of the search for technical solutions for more efficient energy storage, with the adaptation of abandoned coal mine shafts as compressed air tanks, with volumes up to 50,000 m^3 [80].

7.4.2   CAES with low-temperature TES

Low-temperature Adiabatic Compressed Air Energy Storage (LTA-CAES) has rentability advantages and commissioning characteristics in the field of electrical energy storage [81]. In these systems, heat its transferred into useful work [78]: the Carnot relation governs all combustion engines, where heat is the source of energy that is transformed into useful work. This applies to Rankine cycles (steam power plants) or Joule cycles (gas turbines). The temperature range considered for this technology LTA-CAES is 90–200 °C [82].

Some pilot projects applied this LTA-CAES design, such as the 1.5 MW A-CAES demonstration installation in China (Langfang) [83] and a scale-up 10 MW facility in Bijie [84,85], with higher efficiencies than D-CAES [85]. Recent studies [82] suggest a higher efficiency considering CO_2 as the working fluid. This fluid will be analyzed in subsequent sections, hybridized with storage of CO_2.

7.4.3   HTH-CAES system

High-temperature hybrid CAES (HTH-CAES) presents the system configuration that eliminates the combustion emissions required in conventional CAES and mitigates some of the problems encountered in the advanced A-CAES (AA-CAES) [86]. The hybrid HTH-CAES design adds two heating steps through low and high separate TES units. As a result, HTH-CAES operates in smaller volumes and lower pressure, which reduces cost and solves some of the technical problems during expansion. However, there are still some difficulties to choose the correct thermal storage means for the temperature range (500–600 °C) of high-temperature heat of compression [81]. The use of thermal storage in the form of solid materials is proposed [87].

7.4.4   A-CAES coupled with renewable generation technologies

The hybridization of combined heating/cooling systems and power (CHP) systems has been widely touted by researchers to increase energy efficiency and reduce CO_2 emissions [88]. Recently, Diyoke and Wu [89] proposed an ACAES system in hybridization with an energy storage system based on biomass gasification: simultaneously, electricity production and hot water for domestic use. The hybrid system (CAES 1 MW + 0.3 MW dual fuel engine) has an energy efficiency of approximately 38%, while the electrical reaches 30% (the effective is around 34%). The biomass gasification storage system is a technology that can be used locally, where the residual heat can be used as heating or hot water. The system, due to its capacity, can be considered as a micro-CAES system integrating the electricity storage and the heating/cooling system [90].

The integration of both systems (energy storage and cooling/heating) was further studied by Mohammadi *et al.* [91], where the overburden pressure is not higher than 12 bar cavity pressure. The hybridization of both technologies resulted in an energy efficiency of 53.94%. Other configurations of this combination can be found in [36] with the incorporation of a gas engine in the pressurized storage system (considering vessels with a working pressure of less than 5 bar and energy efficiencies of the system of around 50%) or, in [92] with a diesel engine in [93], where it was proposed to use waste heat for biofuel production (syngas) to be used in the expansion stage. Furthermore, researchers look for solutions to improve efficiency through renewable technologies integration, where CAES with solar energy may be mentioned [94].

Finally, the integration of energy storage with anaerobic digestion (BIO-CAES) offers another approach for the heat management, transforming this energy into chemical storage (biogas) [95]. Thus, the process could reach energy efficiencies higher than 70%, with the availability of organic matter being a liming factor for the sizing of the energy storage.

7.4.4.1 Analysis of a hybrid wind–solar energy and CAES system using abandoned infrastructure

The systems mentioned address energy storage systems at lower pressure than conventional CAES. Therefore, the options for underground storage are increased by the fact that not only geological formations are available, but also that obsolete infrastructure can be used. This avoids the search for a suitable geological formation [28,96]. The use of underground space of abandoned mines avoids the need for new excavations, and some surface industrial infrastructures (buildings and facilities) can also house equipment and staff. Thus, the use of abandoned mines to host CAES promotes the development of renewable energies and drastically reduces the amount of investment [97].

Fan *et al.* [97] suggested a WS-CAES hybrid system (hybrid wind and solar system with CAES) that uses old, abandoned coal mine shafts as compressed air energy space. This system has an efficiency of 50.31% and an energy storage density of 3.23 kWh/m^3.

Similar features and projects can be found in [60,98] and projects such as ALACAES [99] or RICAS2020 [100]. Recently, the Spanish SMART project MinEnergy promotes the knowledge base to implement a pilot-scale CAES in abandoned mine drifts.

7.4.5 Compressed gas energy storage system with CO_2 as the working fluid

The storage system CCDS (compressed carbon dioxide storage) is one of the last options considered [101]. Due to the favourable features of heat transfer and capacity to shift into a high-density fluid, the storage with CO_2 in supercritical form (30.98 °C and 7.38 MPa) has been announced. Additionally, the machinery can be simpler (heat exchanger, turbomachinery) [102] and the use of non-ideal gases leads to better thermal efficiency in the Brayton cycle.

It is estimated that a maximum power output value can be obtained, proportional to the compressor output pressure. At the same time, a minimum value for the investment cost can be obtained if the inlet pressure at the turbine is lower than the outlet of the compressor. According to the results, the optimal general energy efficiency is 61.39% [103].

7.5   Conclusions

The future of the electricity generation will involve the generalization of inter-mittent and unmanageable renewable sources: photovoltaic and wind power are the generation sources with the greatest penetration of the last years.

To complete the decarbonization of the electricity system, it will be necessary to incorporate massive energy storage systems. Among the storage options described in this chapter, the use of the subsurface in the form of compressed air storage represents a technically feasible technology. The benchmark of conventional (diabatic) CAES in operation, assuming an energy efficiency of 50%, limits their economic viability. The technological challenge lies in the development of systems for the management of the thermal energy exchanged in the air compression and expansion processes.

In the current context, the success of CAES technologies must inevitably be associated with hybridization with other technologies to increase the efficiency of the overall system and thus its real applicability. This hybridization can come with another energy storage systems such as hydrogen (CAES coupled with hydrogen system, CAHES), or with thermal storage systems such as LTA-CAES or HTH-CAES. Or even with more complex systems, involving technologies from other fields such as biomass gasification (A-CAES + BMGD) or biogas (BIO-CAES) in the field of bio residues utilization. Wind or photovoltaic generation or a system using CO_2 as a working fluid are promising solutions to this hybridization process. Moreover, obsolete infrastructure such as abandoned coal mines may be another hybridization solution to increase techno-economic efficiency of next CAES. Thanks to the hybridizations described here, research on CAES is regaining interest for the use of generated and non-demanded electricity, developing technologies in line with the concept of 'Power to X' (P2X).

CAES solutions should coexist with other mass energy storage solutions with different characteristics, such as pumped hydro or hydrogen, given the countless current and future possibilities for energy storage and their relevance for the expected high penetration of unmanageable renewable energies.

As a conclusion, the overall efficiency of hybridized CAES systems is a viable and feasible response to provide robustness to energy grids in which unmanageable renewable energies are predominantly included.

Acknowledgements

This work is supported by the national SMART MinEnergy project RTC2019-006874-3 financed by the Ministerio de Ciencia e Innovación of Spain. The authors of this work would like to thank Rocío Sánchez Ramos for her contribution as translator, proofreader, and editor of the text.

References

[1]   International Energy Agency. (2020). *Key world energy statistics* (pp. 81). Paris: International Energy Agency.

[2] National Oceanic and Atmospheric Administration (NOAA): https://gml. noaa.gov/ccgg/trends/monthly.html, accessed on 20 September 2020.

[3] Di Pietro, S. (2019). Acuerdo de París: ¿Nuevos compromisos con el medio ambiente o nuevas oportunidades de negocio? [Paris Agreement: New environmental commitments or new business opportunities?] Estudios internacionales (Santiago) *[International Studies (Santiago)]*, 51(192), 57–70.

[4] European Commission. (2020). Stepping up Europe's 2030 climate ambition, COM/2020/562 final. Brussels.

[5] Eurostat database: https://ec.europa.eu/eurostat/web/energy/data/database, accessed on23 September 2020.

[6] Javed, M. S., Ma, T., Jurasz, J., and Amin, M. Y. (2020). Solar and wind power generation systems with pumped hydro storage: Review and future perspectives. *Renewable Energy*, 148, 176–192.

[7] Luo, X., Wang, J., Dooner, M., and Clarke, J. (2015). Overview of current development in electrical energy storage technologies and the application potential in power system operation. *Applied Energy*, 137, 511–536.

[8] Vasudevan, K. R., Ramachandaramurthy, V. K., Venugopal, G., Ekanayake, J. B., and Tiong, S. K. (2021). Variable speed pumped hydro storage: A review of converters, controls and energy management strategies. *Renewable and Sustainable Energy Reviews*, 135, 110156.

[9] Bockris, J. O. M. (2013). The hydrogen economy: Its history. *International Journal of Hydrogen Energy,* 38(6), 2579–2588.

[10] Andrews, J., and Shabani, B. (2012). Re-envisioning the role of hydrogen in a sustainable energy economy. *International Journal of Hydrogen Energy*, 37(2), 1184–1203.

[11] Zhang, F., Zhao, P., Niu, M., and Maddy, J. (2016). The survey of key technologies in hydrogen energy storage. *International Journal of Hydrogen Energy*, 41(33), 14535–14552.

[12] Zhang, Z. (2014). Metal Sulfide Photocatalysts for Hydrogen Generation by Water Splitting Under Illumination of Solar Light (pp. 165–187). Hoboken, NJ: John Wiley & Sons, Inc.

[13] Tarkowski, R. (2019). Underground hydrogen storage: Characteristics and prospects. *Renewable and Sustainable Energy Reviews*, 105, 86–94.

[14] Amrouche, S. O., Rekioua, D., Rekioua, T., and Bacha, S. (2016). Overview of energy storage in renewable energy systems. *International Journal of Hydrogen Energy*, 41(45), 20914–20927.

[15] Kojima, Y. (2019). Hydrogen storage materials for hydrogen and energy carriers. *International Journal of Hydrogen Energy*, 44(33), 18179–18192.

[16] Amez Arenillas, I., Ortega, M. F., García Torrent, J., and Llamas Moya, B. (2021). Hydrogen as an Energy Vector: Present and Future. In Sustaining Tomorrow via Innovative Engineering (pp. 83–129).

[17] Moradi, R., and Groth, K. M. (2019). Hydrogen storage and delivery: Review of the state of the art technologies and risk and reliability analysis. *International Journal of Hydrogen Energy*, 44(23), 12254–12269.

[18]   Abe, J. O., Popoola, A. P. I., Ajenifuja, E., and Popoola, O. M. (2019). Hydrogen energy, economy and storage: review and recommendation. *International Journal of Hydrogen Energy*, 44(29), 15072–15086.

[19]   Stolten, D., and Emonts, B. (Eds.). (2016). Hydrogen Science and Engineering, 2 Volume Set: Materials, Processes, Systems, and Technology (Vol. 1). Weinheim: Wiley-VCH Verlag.

[20]   Lemieux, A., Sharp, K., and Shkarupin, A. (2019). Preliminary assessment of underground hydrogen storage sites in Ontario, Canada. *International Journal of Hydrogen Energy*, 44(29), 15193–15204.

[21]   Sharaf, O. Z., and Orhan, M. F. (2014). An overview of fuel cell technology: Fundamentals and applications. *Renewable and Sustainable Energy Reviews*, 32, 810–853.

[22]   Cheng, C., Blakers, A., Stocks, M., and Lu, B. (2019). Pumped hydro energy storage and 100% renewable electricity for East Asia. *Global Energy Interconnection*, 2(5), 386–392.

[23]   Ichimura, S., and Kimura, S. (2019). Present status of pumped hydro storage operations to mitigate renewable energy fluctuations in Japan. *Global Energy Interconnection*, 2(5), 423–428.

[24]   Vasudevan, K. R., Ramachandaramurthy, V. K., Babu, T. S., and Pouryekta, A. (2020). Synchronverter: A comprehensive review of modifications, stability assessment, applications and future perspectives. *IEEE Access*, 8, 131565–131589.

[25]   Nyeche, E. N., and Diemuodeke, E. O. (2020). Modelling and optimisation of a hybrid PV-wind turbine-pumped hydro storage energy system for mini-grid application in coastline communities. *Journal of Cleaner Production*, 250, 119578.

[26]   Nzotcha, U., Kenfack, J., and Manjia, M. B. (2019). Integrated multi-criteria decision making methodology for pumped hydro-energy storage plant site selection from a sustainable development perspective with an application. *Renewable and Sustainable Energy Reviews*, 112, 930–947.

[27]   Ubertini, S., Facci, A. L., and Andreassi, L. (2017). Hybrid hydrogen and mechanical distributed energy storage. *Energies*, 10(12), 2035.

[28]   Llamas, B., Castañeda, M. D. L. C., Laín, C., and Pous, J. (2017). Multi-criteria algorithm-based methodology used to select suitable domes for compressed air energy storage. *International Journal of Energy Research*, 41(14), 2108–2120.

[29]   Li, L., Weiguo, L., Haojie, L., Jianfeng, Y., and Dusseault, M (2018). Compressed air energy storage: Characteristics, basic principles, and geological considerations. *Advances in Geo-Energy Research*, 2(2), 135–147.

[30]   Han, Z., Guo, S., Wang, S., and Li, W. (2018). Thermodynamic analyses and multi-objective optimization of operation mode of advanced adiabatic compressed air energy storage system. *Energy Conversion and Management*, 174, 45–53.

[31]   Castellani, B., Presciutti, A., Morini, E., Filipponi, M., Nicolini, A., and Rossi, F. (2017). Use of phase change materials during compressed air

expansion for isothermal CAES plants. *Journal of Physics: Conference Series*, 923(1), 012037.

[32] Krawczyk, P., Szabłowski, Ł., Karellas, S., Kakaras, E., and Badyda, K. (2018). Comparative thermodynamic analysis of compressed air and liquid air energy storage systems. *Energy*, 142, 46–54.

[33] Llamas, B., Laín, C., Castañeda, M. C., and Pous, J. (2018). Mini-CAES as a reliable and novel approach to storing renewable energy in salt domes. *Energy*, 144, 482–489.

[34] Salvini, C. (2015). Techno-economic analysis of small size second generation CAES system. *Energy Procedia*, 82, 782–788.

[35] Vollaro, R. D. L., Faga, F., Tallini, A., Cedola, L., and Vallati, A. (2015). Energy and thermodynamical study of a small innovative compressed air energy storage system (micro-CAES). *Energy Procedia*, 82, 645–651.

[36] Yao, E., Wang, H., Wang, L., Xi, G., and Maréchal, F. (2016). Thermo-economic optimization of a combined cooling, heating and power system based on small-scale compressed air energy storage. *Energy Conversion and Management*, 118, 377–386.

[37] Tola, V., Meloni, V., Spadaccini, F., and Cau, G. (2017). Performance assessment of adiabatic compressed air energy storage (A-CAES) power plants integrated with packed-bed thermocline storage systems. *Energy Conversion and Management*, 151, 343–356.

[38] Ruoso, A. C., Caetano, N. R., and Rocha, L. A. O. (2019). Storage gravitational energy for small scale industrial and residential applications. *Inventions*, 4(4), 64.

[39] Tong, Z., Cheng, Z., and Tong, S. (2020). A review on the development of compressed air energy storage in China: Technical and economic challenges to commercialization. *Renewable and Sustainable Energy Reviews*, 135, 110178.

[40] Albawab, M., Ghenai, C., Bettayeb, M., and Janajreh, I. (2020). Sustainability performance index for ranking energy storage technologies using multi-criteria decision-making model and hybrid computational method. *Journal of Energy Storage*, 32, 101820.

[41] Mongird, K., Viswanathan, V., Balducci, P., *et al.* (2020). An evaluation of energy storage cost and performance characteristics. *Energies*, 13(13), 3307.

[42] Wang, J., Lu, K., Ma, L., *et al.* (2017). Overview of compressed air energy storage and technology development. *Energies*, 10(7), 991.

[43] Zhou, Q., Du, D., Lu, C., He, Q., and Liu, W. (2019). A review of thermal energy storage in compressed air energy storage system. *Energy*, 188, 115993.

[44] Venkataramani, G., Parankusam, P., Ramalingam, V., and Wang, J. (2016). A review on compressed air energy storage – A pathway for smart grid and polygeneration. *Renewable and Sustainable Energy Reviews*, 62, 895–907.

[45] Budt, M., Wolf, D., Span, R., and Yan, J. (2016). A review on compressed air energy storage: Basic principles, past milestones and recent developments. *Applied Energy*, 170, 250–268.

[46] Apex-CAES. (2021). http://www.apexcaes.com/project, accessed on 2 February 2021.

[47] de Biasi, V. (2009). Fundamental analyses to optimize adiabatic CAES plant efficiencies. *Gas Turbine World*, 26–28.

[48] Jubeh, N. M., and Najjar, Y. S. (2012). Green solution for power generation by adoption of adiabatic CAES system. *Applied Thermal Engineering*, 44, 85–89.

[49] Peng, H., Yang, Y., Li, R., and Ling, X. (2016). Thermodynamic analysis of an improved adiabatic compressed air energy storage system. *Applied Energy*, 183, 1361–1373.

[50] Kosamana, B., and Venkatachalam, K. K. (2020). U.S. Patent No. 10,584,634. Washington, DC: U.S. Patent and Trademark Office.

[51] Bieber, M., Marquardt, R., and Moser, P. (2010). The ADELE Project: Development of an adiabatic CAES plant towards marketability. In *5th International Renewable Energy Storage Conference*, General Electric, Bonn, München (Germany).

[52] Wojcik, J. D., and Wang, J. (2018). Feasibility study of combined cycle gas turbine (CCGT) power plant integration with adiabatic compressed air energy storage (ACAES). *Applied Energy*, 221, 477–489.

[53] Li, Y., Miao, S., Zhang, S., *et al.* (2019). A reserve capacity model of AA-CAES for power system optimal joint energy and reserve scheduling. *International Journal of Electrical Power & Energy Systems*, 104, 279–290.

[54] Dinker, A., Agarwal, M., and Agarwal, G. D. (2017). Heat storage materials, geometry and applications: A review. *Journal of the Energy Institute*, 90(1), 1–11.

[55] Hadorn, J. C. (2008, October). Advanced storage concepts for active solar energy – IEA SHC Task 32 2003-2007. In *Eurosun 1st International Conference on Solar Heating, Cooling and Buildings*, Lisbon (pp. 1–8).

[56] Atalay, H. (2019). Performance analysis of a solar dryer integrated with the packed bed thermal energy storage (TES) system. *Energy*, 172, 1037–1052.

[57] Ndukwu, M. C., Bennamoun, L., Abam, F. I., Eke, A. B., and Ukoha, D. (2017). Energy and exergy analysis of a solar dryer integrated with sodium sulfate decahydrate and sodium chloride as thermal storage medium. *Renewable Energy*, 113, 1182–1192.

[58] Nordell, B. (2013). Underground thermal energy storage (UTES). In *International Conference on Energy Storage: 16/05/2012-18/05/2012.*

[59] Nordell, B. (1994). Borehole heat store design optimization, PhD thesis 1994:137D. Luleå University of Technology, Sweden, pp. 250.

[60] Kim, H. M., Rutqvist, J., Ryu, D. W., Choi, B. H., Sunwoo, C., and Song, W. K. (2012). Exploring the concept of compressed air energy storage (CAES) in lined rock caverns at shallow depth: A modeling study of air tightness and energy balance. *Applied Energy*, 92, 653–667.

[61] Saputro, E. A., and Farid, M. M. (2018). A novel approach of heat recovery system in compressed air energy storage (CAES). *Energy Conversion and Management*, 178, 217–225.

[62] Pickard, W. F., Hansing, N. J., and Shen, A. Q. (2009). Can large-scale advanced-adiabatic compressed air energy storage be justified economically in an age of sustainable energy? *Journal of Renewable and Sustainable Energy*, 1(3), 033102.

[63] Mozayeni, H., Negnevitsky, M., Wang, X., Cao, F., and Peng, X. (2017). Performance study of an advanced adiabatic compressed air energy storage system. *Energy Procedia*, 110, 71–76.

[64] Ortega-Fernández, I., Zavattoni, S. A., Rodríguez-Aseguinolaza, J., D'Aguanno, B., and Barbato, M. C. (2017). Analysis of an integrated packed bed thermal energy storage system for heat recovery in compressed air energy storage technology. *Applied Energy*, 205, 280–293.

[65] Castellani, B., Presciutti, A., Filipponi, M., Nicolini, A., and Rossi, F. (2015). Experimental investigation on the effect of phase change materials on compressed air expansion in CAES plants. *Sustainability*, 7, 9773–9786.

[66] Zhang, X., Xu, Y., Zhou, X., *et al.* (2018). A near-isothermal expander for isothermal compressed air energy storage system. *Applied Energy*, 225, 955–964.

[67] Fu, H., Jiang, T., Cui, Y., and Li, B. (2019). Design and operational strategy research for temperature control systems of isothermal compressed air energy storage power plants. *Journal of Thermal Science*, 28(2), 204–217.

[68] Olabi, A. G., Wilberforce, T., Ramadan, M., Abdelkareem, M. A., and Alami, A. H. (2021). Compressed air energy storage systems: Components and operating parameters–A review. *Journal of Energy Storage*, 34, 102000.

[69] Chen, H., Peng, Y. H., Wang, Y. L., and Zhang, J. (2020). Thermodynamic analysis of an open type isothermal compressed air energy storage system based on hydraulic pump/turbine and spray cooling. *Energy Conversion and Management*, 204, 112293.

[70] Kantharaj, B., Garvey, S., and Pimm, A. (2015). Compressed air energy storage with liquid air capacity extension. *Applied Energy*, 157, 152–164.

[71] Morgan, R., Nelmes, S., Gibson, E., and Brett, G. (2015). Liquid air energy storage – Analysis and first results from a pilot scale demonstration plant. *Applied Energy*, 137, 845–853.

[72] Ding, Y., Tong, L., Zhang, P., Li, Y., Radcliffe, J., and Wang, L. (2016). Liquid air energy storage. In *Storing Energy*, T. Letcher (Ed.) (pp. 167–181). Oxford: Elsevier.

[73] Sciacovelli, A., Vecchi, A., and Ding, Y. (2017). Liquid air energy storage (LAES) with packed bed cold thermal storage – From component to system level performance through dynamic modelling. *Applied Energy*, 190, 84–98.

[74] Ameel, B., T'Joen., C., De Kerpel, K., *et al.* (2013) Thermodynamic analysis of energy storage with a liquid air Rankine cycle. *Applied Thermal Engineering*, 52(1), 130–140.

[75] Liu, J., Xia, H., Chen, H., Tan, C., and Xu, Y. (2010). A novel energy storage technology based on liquid air and its application in wind power. *Journal of Engineering Thermophysics*, 31, 1993–1996.

[76] Antonelli, M., Barsali, S., Desideri, U., Giglioli, R., Paganucci, F., and Pasini, G. (2017). Liquid air energy storage: Potential and challenges of hybrid power plants. *Applied Energy*, 194, 522–529.

[77] Bartela, Ł. (2020). A hybrid energy storage system using compressed air and hydrogen as the energy carrier. *Energy*, 196, 117088.

[78]   Wolf, D., and Budt, M. (2014). LTA-CAES – A low-temperature approach to adiabatic compressed air energy storage. *Applied Energy*, 125, 158–164.

[79]   Szablowski, L., Krawczyk, P., Badyda, K., Karellas, S., Kakaras, E., and Bujalski, W. (2017). Energy and exergy analysis of adiabatic compressed air energy storage system. *Energy*, 138, 12–18.

[80]   Lutyński, M., Smolnik, G., and Waniczek, S. (2019, May). Underground coal mine workings as potential places for compressed air energy storage. In *IOP Conference Series: Materials Science and Engineering* (Vol. 545, No. 1, p. 012014). IOP Publishing.

[81]   Guo, C., Xu, Y., Guo, H., *et al.* (2019). Comprehensive exergy analysis of the dynamic process of compressed air energy storage system with low-temperature thermal energy storage. *Applied Thermal Engineering*, 147, 684–693.

[82]   Zhang, Y., Yao, E., and Wang, T. (2021). Comparative analysis of compressed carbon dioxide energy storage system and compressed air energy storage system under low-temperature conditions based on conventional and advanced exergy methods. *Journal of Energy Storage*, 35, 102274.

[83]   Chinese Academy of Sciences, Institute of Engineering Thermophysics, annual report 2015. http://www.iet.cn/cxwh/nb/201508/P020151112332330 249036.pdf, accessed on 1 February 2021.

[84]   Chinese Academy of Sciences: http://en.cnesa.org/latest-news/2019/9/29/ compressed-air-energy-storage-becoming-a-leading-energy-storage-technol-ogy, accessed on 1 February 2021.

[85]   King, M., Jain, A., Bhakar, R., Mathur, J., and Wang, J. (2021). Overview of current compressed air energy storage projects and analysis of the potential underground storage capacity in India and the UK. *Renewable and Sustainable Energy Reviews*, 139, 110705.

[86]   Houssainy, S., Janbozorgi, M., Ip, P., and Kavehpour, P. (2018). Thermodynamic analysis of a high temperature hybrid compressed air energy storage (HTH-CAES) system. *Renewable Energy*, 115, 1043–1054.

[87]   Baghaei Lakeh, R., Villazana, I. C., Houssainy, S., Anderson, K. R., and Kavehpour, H. P. (2016). Design of a modular solid-based thermal energy storage for a hybrid compressed air energy storage system. *Proceedings of the ASME 2016 10th International Conference on Energy Sustainability collocated with the ASME 2016 Power Conference and the ASME 2016 14th International Conference on Fuel Cell Science, Engineering and Technology*. Volume 2: ASME 2016 Energy Storage Forum. Charlotte, North Carolina, USA. 26–30 June 2016. V002T01A008. ASME.

[88]   Liu, M., Shi, Y., and Fang, F. (2014). Combined cooling, heating and power systems: A survey. *Renewable and Sustainable Energy Reviews*, 35, 1–22.

[89]   Diyoke, C., and Wu, C. (2020). Thermodynamic analysis of hybrid adiabatic compressed air energy storage system and biomass gasification storage (A-CAES+ BMGS) power system. *Fuel*, 271, 117572.

[90]   Kim, Y. M., and Favrat, D. (2010). Energy and exergy analysis of a micro-compressed air energy storage and air cycle heating and cooling system. *Energy*, 35(1), 213–220.

[91] Mohammadi, A., Ahmadi, M. H., Bidi, M., Joda, F., Valero, A., and Uson, S. (2017). Exergy analysis of a Combined Cooling, Heating and Power system integrated with wind turbine and compressed air energy storage system. Energy Conversion and Management, 131, 69–78.

[92] Zhang, X., Chen, H., Xu, Y., *et al.* (2017). Distributed generation with energy storage systems: A case study. *Applied Energy*, 204, 1251–1263.

[93] Denholm, P. (2006). Improving the technical, environmental and social performance of wind energy systems using biomass-based energy storage. *Renewable Energy*, 31(9), 1355–1370.

[94] Garrison, J. B., and Webber, M. E. (2011). An integrated energy storage scheme for a dispatchable solar and wind powered energy system and analysis of dynamic parameters. In *ASME 2011 5th International Conference on Energy Sustainability*, Vol. 54686, pp. 2101–2111.

[95] Llamas, B., Ortega, M. F., Barthelemy, G., de Godos, I., and Acién, F. G. (2020). Development of an efficient and sustainable energy storage system by hybridization of compressed air and biogas technologies (BIO-CAES). *Energy Conversion and Management*, 210, 112695.

[96] Jannelli, E., Minutillo, M., Lavadera, A. L., and Falcucci, G. (2014). A small-scale CAES (compressed air energy storage) system for stand-alone renewable energy power plant for a radio base station: A sizing-design methodology. *Energy*, 78, 313–322.

[97] Fan, J., Liu, W., Jiang, D., *et al.* (2018). Thermodynamic and applicability analysis of a hybrid CAES system using abandoned coal mine in China. *Energy*, 157: 31–44.

[98] Rutqvist, J., Kim, H. M., Ryu, D. W., Synn, J. H., and Song, W. K. (2012). Modeling of coupled thermodynamic and geomechanical performance of underground compressed air energy storage in lined rock caverns. *International Journal of Rock Mechanics and Mining Sciences*, 52, 71–81.

[99] Zavattoni, S., Roncolato, J., Geissbühler, L., *et al.* (2019). The Swiss AA-CAES pilot plant: CFD modeling and validation of the TES system. In *35th International CAE Conference and Exhibition 2019. CAE Conference.*

[100] Buckstegge, F., Michel, T., Zimmermann, M., Roth, S., and Schmidt, M. (2016). Advanced rock drilling technologies using high laser power. *Physics Procedia*, 83, 336–343.

[101] Chen, L. X., Xie, M. N., Zhao, P. P., Wang, F. X., Hu, P., and Wang, D. X. (2018). A novel isobaric adiabatic compressed air energy storage (IA-CAES) system on the base of volatile fluid. *Applied Energy*, 210, 198–210.

[102] Akikur, R. K., Saidur, R., Ping, H. W., and Ullah, K. R. (2014). Performance analysis of a co-generation system using solar energy and SOFC technology. *Energy Conversion and Management*, 79, 415–430.

[103] Liu, Z., Yang, X., Jia, W., Li, H., and Yang, X. (2020). Justification of CO_2 as the working fluid for a compressed gas energy storage system: A thermodynamic and economic study. *Journal of Energy Storage*, 27, 101132.

Chapter 8

Compressed air energy storage system dynamic modelling and simulation

Jieren Ke[1], Wei He[1], Mark Dooner[1], Xing Luo[1] and Jihong Wang[1]

The compressed air energy storage (CAES) system is a very complex system with multi-time-scale physical processes. Following the development of computational technologies, research on CAES system model simulation is becoming more and more important for resolving challenges in system pre-design, optimization, control and implementation. In this chapter, five types of simulation model for CAES system and components have been explained and compared based on the discharging process of the CAES. Principles for choosing suitable model methods targeting different purposes for CAES system have been described, and a novel data-driven dynamic simulation approach for the complex system is demonstrated. The result shows that the data-driven simulation approach can reduce computational cost sharply and may help build CAES system-level real-time simulation in the future.

Keywords: Compressed air energy storage; multi-physical system simulation; data-driven modelling; model selection

8.1 Introduction

The idea of storing electrical energy via compressed air can be traced back to the early 1940s [1]. In 1978, the world's first CAES power plant, Huntorf, was commissioned and is still in operation [2]. Since then, the exploration of the approaches for achieving higher efficiency and relaxing the limitations of CAES has been continued. A simplified modern CAES system diagram is shown in Figure 8.1.

In recent years, increasing intermittent renewable energy integration has led to an increase in demand for electrical energy storage and grid-scale energy storage in particular. This has accelerated the research activities on CAES technology. The scale and complexity of CAES systems mean that modelling and simulation studies

---

[1]Power and Control Research Laboratory, University of Warwick, Coventry, UK

Power grid

Power input   Power output

Compressor (1) Drive Motor (6) Generator Drive (7) Expander (3) Exhausted air

Inlet air

Compressed high-temperature air

HEXs and TES (2)

HEXs and TES or fuel combustion (4)

Compressed high-temperature air

Compressed low-temperature air

Compressed low-temperature air

Air reservoir (5)

Charging process

Discharging process

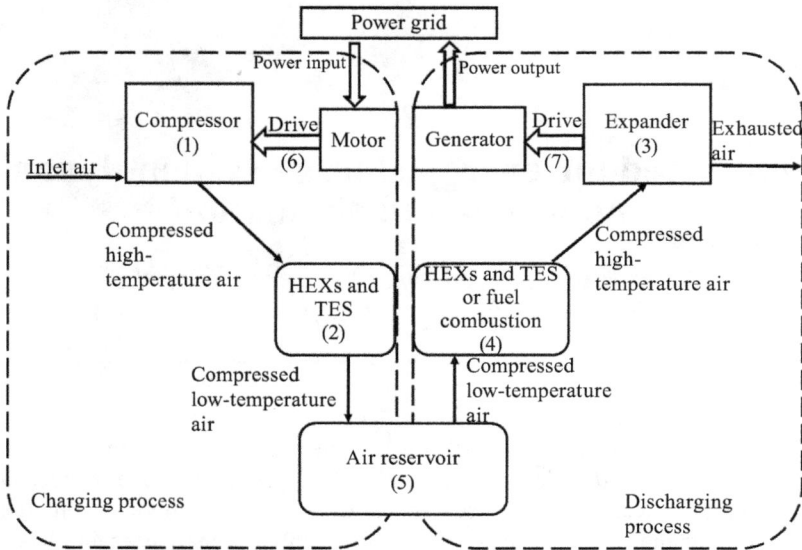

Figure 8.1   A simplified CAES system scheme

Table 8.1   Comparison of the number of articles in the CAES system model simulation area

Keyword	Before 1990	1990–2000	2000–2010	2010–Present
CAES system	13,000	6,080	16,300	17,200
CAES system model simulation	773	952	2,310	9,700

are an important tool to analyse and evaluate innovative design ideas and practical project feasibility. The trend of escalation of published literature in CAES modelling and simulation is evidenced in the figures shown in Table 8.1 compiled from a search of the Google Scholar system. Over 50% of the overall articles concerning CAES have been published since 2010.

CAES is a complex system which involves multi-time-scale physical processes (e.g. thermal, mechanical, electrical and fluid dynamics). It is almost impossible to perform physical system-level tests before a CAES plant is fully built and it is challenging to pre-design optimal control strategies. Therefore, dynamic modelling and simulation are essential for achieving a comprehensive understanding of the system behaviours and improving designs, which may lead to safety enhancement and cost reduction. Additionally, the whole-system dynamic modelling can help with the development of control strategies.

This chapter presents the recent work conducted at the University of Warwick in CAES dynamic modelling and simulation, which covers individual key system component modelling and system-level modelling. Components for discharging process dynamic modelling of grid-scale CAES is illustrated in detail. To improve the simulation speed and increase the robustness of system-level simulation, a new combined data-driven and analytic modelling approach is introduced. Guidance for choosing suitable model type for different simulation purpose is presented afterwards, and an example of data-driven and lumped-parameter mixed model dynamic simulation for a large CAES discharging process is shown at last.

8.2 CAES discharging system analysis, simulation and modelling

To implement CAES system simulation, models for components should be developed first. Limited by space, first-principle models (FPMs) for each key component in large-scale CAES system discharging process are described.

For modelling the CAES system, five types of modelling methods exist for different purposes:

1. Thermodynamic models. This modelling process is normally based on the ideal gas theory, and assuming the target thermodynamic process is isentropic, isothermal or polytropic. Some thermodynamic models consider realistic gas in the simulation, which can improve their accuracy. Some software, e.g. REFPROP and CoolProp, allow users to use real gas properties in the modelling process. These models are widely utilized when analysing system/component characteristics under steady states. Some examples can be found in the literature [3–10] for compressors, expanders, thermal energy storages (TESs), heat exchangers (HEXs) and high-pressure air reservoirs.

2. Empirical models. Empirical models are derived purely from experimental data and have been widely used at system-level because of the low computation. This kind of model cannot reveal the physical relationship between the component parameters and its performance. Once the parameters of the component have been changed, the model cannot further predict the component performance accurately. In most case, it is a map between the input and output experimental data.

3. Data-driven models (DDMs). DDMs are similar to empirical models, yet they are derived from target component input/output data. The data can either from experiment or verified simulation. Like empirical models, they cannot reveal the physical relationship between the component parameters and its performance. Unlike empirical models, they can represent dynamic behaviours of the target component, as the derived DDM are not map between the input and output data, but more an approximation function between two time-series sequences.

4. Lumped-parameter models. They are developed based on geometric parameters of the target component and first principles, e.g. laws of mass, momentum and energy to simulate how variables change during target physical processes in the component. This kind of model uses differential equation sets

to link the dynamic behaviours and geometric parameters of the component. They imply some assumptions; for example, the air temperature within a compressor and the pressure within the air reservoir is the same throughout their volume. Compared to empirical models, they can reveal a clear physical-based relationship between the component's performance and its parameters. To gain this advantage, they require larger computation cost than empirical models.

5. Distributed-parameter models or full geometry-based models. They are also developed based on the geometric parameters of the target component and first principles. The difference between distributed-parameter models and lumped-parameter models is that distributed-parameter models describe physical processes using partial differential equations along three spatial dimensions of the component; thus, they normally require higher computational resources than lumped-parameter models. A typical example is computational fluid dynamics (CFD). CFD can provide a fully resolved model in space and is a powerful approach for simulating the expander performance accurately. However, CFD models are normally in three dimensions and they are considered as high-fidelity models, which requires extremely intensive computation which will take a much longer time to solve, compared to other kinds of models. As a result, it is not suitable for (real-time) dynamic control purposes and it is difficult to implement in embedded systems for system estimation. It is mainly used for the analysis and optimization of a well-designed component, e.g., an expander. However, some parameter-distributed models are not 3D, but 1D or 2D, e.g. some TES and HEX models. They often trade more detailed system properties change along one or two dimensions for less computation cost.

Key system component modelling and system-level modelling are the focus for the work presented here; thus, thermodynamic models, DDMs and lumped models are considered for achieving the purpose. The 3D distributed-parameter models are not included due to its high computational request, while 1D or 2D may be considered based on its computational cost. Except for DDM and empirical models, other types of models are built from first principles of the component. They can also be called as FPM.

8.2.1 *Air expander*

An air expander is a crucial component in a CAES system. It converts the exergy of the pressurized air into mechanical power to drive the generator. It determines the rated power output and contributes to the overall energy conversion efficiency. Two types of expanders are commonly used in CAES systems: the volume types, i.e. screw and scroll expanders, reciprocating expanders, and velocity types, i.e. radial and axial turbines. The former type operates in an intermittent mode which is naturally cyclic, while the latter type operates continuously without interruption of the flow at any moment during expansion. Velocity type expanders are also called turbines. They convert energy between the rotational kinetic energy of the impeller and the working fluid's dynamic energy, while volume type machines use chambers with variable dimensions to complete compression or expansion processes.

The physical structure and operation principles of a typical radial expander can be found in [11]. Based on the mean-line model, illustrated velocity triangles of air at the inlet and outlet of the stator and rotor are shown in Figure 8.2. These turbines can be regarded as a purely resistive fluid flow component in which the accumulations of mass, momentum and energy inside the turbine flow path are negligible [12]. Therefore, the unsteadiness of internal flow is not considered in the modelling.

The changes in the air's momentum result in the work released by the turbine, which drives an externally applied torque, namely the generator in the CAES. The specific work can be described as follows:

$$w_t = \Delta h = u_2 c_{y2} - u_3 c_{y3} \tag{8.1}$$

where subscripts 2 and 3 refer to the inlet and the outlet, respectively, u and c_y are the tangential velocities of the rotor and air, respectively, and Δh refers to the energy head of air. When a 90-degree inward-flow radial (IFR) turbine is considered at its nominal design condition, it leads to $c_{y3} = 0$ and $c_{y2} = u_2$. Furthermore, the tangential velocity of the rotor is dependent on the rotor size and the rotation speed, which is

$$u_2 = \frac{\pi N D_2}{60} \tag{8.2}$$

where D_2 is the rotor diameter, and N its rotational speed in rpm. Within the stator, the energy converts from the pressure head to the velocity head. The velocity of the air at the outlet of the stator can be approximated by assuming adiabatic airflow in an equivalent nozzle with energy loss:

$$c_2 = \sqrt{2(\Delta h_{1-2s} - \Delta h_{2loss})} = \sqrt{2\left(\frac{\kappa}{\kappa - 1} RT_1\left[\left(\frac{p_2}{p_1}\right)^{\frac{\kappa-1}{\kappa}} - 1\right] - \Delta h_{2loss}\right)} \tag{8.3}$$

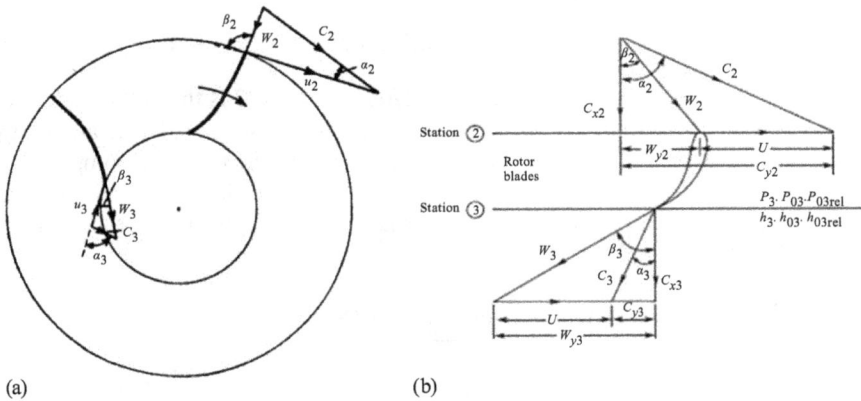

(a)   (b)

Figure 8.2   Schematic velocity triangles at the inlet and outlet of the rotor [13]

where c_2 is the absolute velocity, p is the pressure, R is the gas constant, T is the temperature, κ is the isentropic index, subscripts 1 and 2 denote the inlet and outlet of the stator, respectively, Δh_{1-2s} is the enthalpy of the air and Δh_{2loss} is the enthalpy loss in the stator. It can be approximated by the following [14]:

$$\Delta h_{2loss} = \frac{1}{2}\xi_2 c_2^2 \tag{8.4}$$

in which ξ_2 is the enthalpy loss coefficient.

In off-design operations, a sudden change of the flow direction is accompanied by a shock formation, which is the dissipation of the kinetic energy, which transfers heat to the air at the inlet of the rotor. Based on the assumed same pressure of the air at the stator's outlet and the rotor's inlet, equations of mass and energy conservation can be derived. The energy balance becomes as follows [15]:

$$C_p T_2 + \frac{c_2^2}{2} = C_p T_2' + \frac{c_2'^2}{2} + [c_2 \cos \alpha_2 - c_2 \sin \alpha_2 \frac{T_2'}{T_2} \cot \beta_2 - u_2]u_2 \tag{8.5}$$

where T_2' and c_2' are, respectively, the varied temperature and velocity due to the sudden deflection of the flow, C_p the specific heat capacity, α_2 the rotor absolute inlet angle and β_2 is the rotor relative inlet. The equation of continuity in a radial direction is

$$\frac{w_2' \sin \beta_2}{c_2 \sin \alpha_2} = \frac{T_2'}{T_2} \tag{8.6}$$

where w_2' is the varied relative velocity. Based on (8.5) and (8.6), changing temperature T_2' due to incidence loss can be obtained. Furthermore, the relative velocity of air at the outlet of the rotor is

$$w_3^2 = w_2^2 + 2(\Delta h_{2-3s} - \Delta h_{3loss}) + u_3^2 - u_2^2 \tag{8.7}$$

where Δh_{2-3s} is the enthalpy of the air and $\Delta h_{3loss} = \xi_3 w_3^2/2$ is the enthalpy loss of the energy conversion between air and rotor [14]. Therefore, considering all energy losses, the torque produced in the rotor can be obtained. The torque is composed of two parts, namely, that due to the sudden deflection of the flow when entering the rotor, and that produced within the rotor passage. Therefore, the two torques can be expressed as follows [14]:

$$\tau_l = \frac{[c_2 \cos \alpha_2 - c_2 \sin \alpha_2 \frac{T_2'}{T_2} \cot \beta_2 - u_2]D_2}{2};$$

$$\tau_r = \frac{[c_2' \cos \alpha_2' + \frac{D_3}{D_2} c_3 \cos \alpha_3]D_2}{2} \tag{8.8}$$

where τ_s is the torque due to flow deflection and τ_r is the mechanical torque of the rotor. Accordingly, the isentropic efficiency of the turbine is

$$\eta_s = \frac{(\tau_s + \tau_r)\omega}{\Delta h_{13s}} = \frac{(\tau_s + \tau_r)}{\frac{\kappa}{\kappa-1}RT_1\left[\left(\frac{p_3}{p_1}\right)^{\frac{\kappa-1}{\kappa}} - 1\right]} \frac{2\pi N}{60} \tag{8.9}$$

where ω is the rotation speed of the turbine in rad/s.

To ensure the mass balance, it needs to find an appropriate pressure P_2 to satisfy the conservation of energy as shown in (8.5) and the conservation of mass $\dot{m}_N = \dot{m}_R$, which are shown as follows:

$$\dot{m}_N = c_2 \sin \alpha_2 \pi D_2 b_2 \rho_1 \left(\frac{P_2}{P_1}\right)^{1/k}; \dot{m}_R = c_3 \sin \alpha_3 \pi D_3 b_3 \rho_2 \frac{T_2}{T_2'}\left(\frac{P_3}{P_2}\right)^{1/k} \tag{8.10}$$

Therefore, if the two mass flow rates are not equivalent, a repeat calculation must be made to find the appropriate pressure ratios through the stator and rotor, until the mass balance is achieved.

8.2.2 HEX and TES

TES systems have been used jointly with CAES plants to recycle the exhaust thermal energy from the compression process [16]. It is a key component in CAES systems, which absorbs the heat from the compression cycle and releases it in the expansion cycle. HEXs usually are bridges between the TES and CAES compressors/expanders for cooling down or heating the air. In [17], it is clear that the isentropic efficiencies of compressors and turbines and the heat transfer rates of HEXs have a dominant impact on the whole CAES system efficiency.

In the presented work, several types of TES are modelled, including packed-bed TES (PBTES) with phase change material (PCM), PBTES with solid sensible heat storage, and water tanks. For HEXs, there are both a steady-state model and a dynamic model. Some models are explained in [18]; in this section, the PBTES PCM model is explained.

During the charging process, the increasing heat stored in TES can be expressed based on the thermal balance:

$$\begin{cases} \text{sensible}: h_{\text{sto}}^{\text{sen}} A_{\text{sto}}\left(T_{\text{gas}}^{\text{in}} - T_{\text{mod}}\right) = \left(m_{\text{pcm}}c_{\text{pcm}} + m_{\text{tube}}c_{\text{tube}}\right)\frac{\mathrm{d}T_{\text{mod}}}{\mathrm{d}t} \\ \text{latent}: h_{\text{sto}}^{\text{lat}} A_{\text{sto}}\left(T_{\text{gas}}^{\text{in}} - T_{\text{mod}}\right) = \left(m_{\text{pcm}}c_{\text{pcm}} + m_{\text{tube}}c_{\text{tube}} + m_{\text{pcm}}c_{\text{h}}\right)\frac{\mathrm{d}T_{\text{mod}}}{\mathrm{d}t} \end{cases} \tag{8.11}$$

where $h_{\text{sto}}^{\text{sen}}$ and $h_{\text{sto}}^{\text{lat}}$ refer to the average convective heat transfer coefficients of the sensible and latent thermal storage modules, respectively; A_{sto} refers to the heat transfer area of the module in the thermal storage process; $T_{\text{gas}}^{\text{in}}$ and T_{mod} refer to the temperature of the gas inlet of the module and the mean temperatures of the module

(PCMs and tube bundles), respectively; m_{pcm} and c_{pcm} refer to the mass and specific heat capacity of the PCMs, respectively; m_{tube} and c_{tube} refer to the mass and specific heat capacity of the steel tube, respectively; and c_h refers to the specific heat capacity of the equivalent latent heat of the PCMs.

The following equation is also derived from the energy balance:

$$\left(Q_{pcm}^{sto} + Q_{tube}^{sto}\right) = m_{gas}c_{gas}\left(T_{gas}^{in} - T_{gas}^{out}\right)\eta_{sto} \tag{8.12}$$

where Q_{pcm}^{sto} and Q_{tube}^{sto} refer to the heat stored in the PCMs and the tubes, respectively; η_{sto} refers to the internal thermal storage coefficient of the module; and T_{gas}^{out} is the gas temperature in the outlet of the module.

During the thermal discharging process, the air absorbs the heat stored in the module via the convection heat transfer. Following the thermal balance, the thermal release process of the module is as follows:

$$\begin{cases} \text{Sensible}: -h_{rel}^{sen}A_{rel}\left(T_{mod} - T_{gas}^{in}\right) = \left(m_{pcm}c_{pcm} + m_{tube}c_{tube}\right)\dfrac{dT_{mod}}{dt} \\ \text{Latent}: -h_{rel}^{lat}A_{rel}\left(T_{mod} - T_{gas}^{in}\right) = \left(m_{pcm}c_{pcm} + m_{tube}c_{tube} + m_{pcm}c_h\right)\dfrac{dT_{mod}}{dt} \end{cases} \tag{8.13}$$

where h_{rel}^{sen} and h_{rel}^{lat} denote the mean convective heat transfer coefficients of the sensible and latent thermal modules, respectively, and A_{rel} denotes the heat transfer area of the module in the thermal release process.

The following equation is derived from energy balance:

$$\left(Q_{pcm}^{rel} + Q_{tube}^{rel}\right) = m_{gas}c_{gas}\left(T_{gas}^{out} - T_{gas}^{in}\right)\eta_{rel} \tag{8.14}$$

where Q_{pcm}^{rel} and Q_{tube}^{rel} denote the heat released by the PCMs and the tubes, respectively, and η_{rel} refers to the internal thermal release coefficient of the module.

The dynamic models of thermal storage modules introduced above are limited to modules with an incompressible flow in which Reynolds number is relatively low.

8.2.3   Generator

A generator in a CAES system is co-shafted with the expander and converts mechanical power delivered by the expander to electrical power. Depending on the power rating of a CAES system, different generators can be chosen. Some small-scale CAES systems with a capacity of several kW are equipped with permanent magnet synchronous generators [19,20], whereas some smaller CAES systems use permanent magnet (PM) DC generators [21,22]. In [23], a 15 MW CAES system uses a three-phase asynchronous generator. In Huntorf, a 290 MW synchronous generator is adopted [2]. McIntosh plant built in 1991 also uses synchronous generator [24].

Compared to a PM synchronous machine, the non-PM synchronous machine is more flexible. The power factor of the non-PM machine can be controlled by changing the excitation current. In this case, it is better suited for a CAES system, as it requires flexibility to handle variable demand.

The equivalent circuits of the synchronous machine represented in the d-q-0 frame are shown in Figure 8.3, which take into account the dynamics of armature, field and damper windings.

The model is established in the per-unit system. The methods of selecting base values are given as

$$\begin{cases} U_b = U_m = \sqrt{2}U_N \\ I_b = I_m = \sqrt{2}I_N \\ t_b = \dfrac{1}{2\pi f_N} \end{cases} \tag{8.15}$$

where U_b, I_b and t_b are the base values of voltage, current and time, respectively; U_m and I_m are the amplitude of the phase voltage and phase current, respectively; U_N and I_N are the nominal voltage and nominal current, respectively; and f_N represents the nominal frequency.

The flux equations are described as follows:

$$\begin{cases} \psi_d = -L_d i_d + L_{md}(i_f + i_D) \\ \psi_q = -L_q i_q + L_{mq} i_Q \\ \psi_f = L_f i_f + L_{md}(i_D - i_d) \\ \psi_D = L_D i_D + L_{md}(i_f - i_d) \\ \psi_Q = L_Q i_Q - L_{mq} i_q \end{cases} \tag{8.16}$$

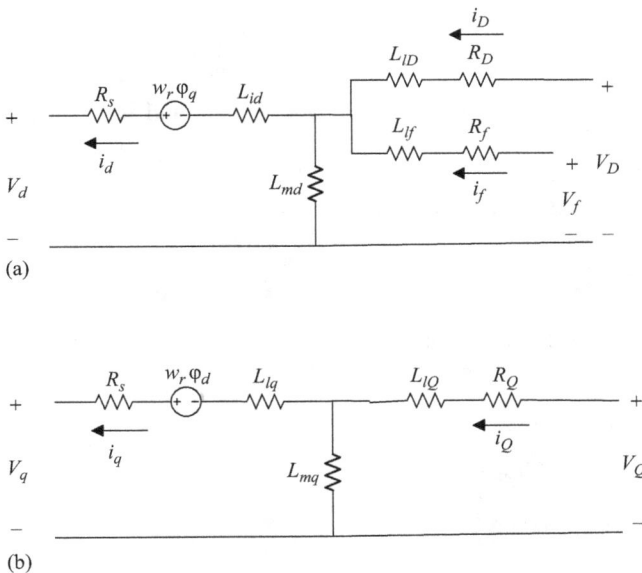

Figure 8.3 The equivalent circuit of synchronous machine (dynamic): (a) d-axis; (b) q-axis

where ψ stands for flux, L stands for inductance, and i stands for current. The subscripts used are defined as followed: d and q represent d-axis and q-axis quantity, respectively, m represents magnetizing inductance, f, D and Q represent the field, d-axis damper and q-axis damper windings quantity, respectively.

Besides the d, q axis armature windings inductance, the field windings inductance and the d, q axis damper windings inductance are related to d-axis magnetizing inductance and q-axis magnetizing inductance. The relationship can be described as follows:

$$\begin{cases} L_q = L_l + L_{mq} \\ L_d = L_l + L_{md} \\ L_Q = L_{lQ} + L_{mq} \\ L_f = L_{lf} + L_{md} \\ L_D = L_{lD} + L_{md} \end{cases} \tag{8.17}$$

where the subscript l stands for leakage inductance.

The voltage equations are described as follows:

$$\begin{cases} U_d = -R_s i_d - w_r \psi_q + \dfrac{d\psi_d}{dt} \\ U_q = -R_s i_q + w_r \psi_d + \dfrac{d\psi_q}{dt} \\ U_f = R_f i_f + \dfrac{d\psi_f}{dt} \\ U_D = R_D i_D + \dfrac{d\psi_D}{dt} \\ U_Q = R_Q i_Q + \dfrac{d\psi_Q}{dt} \end{cases} \tag{8.18}$$

where R stands for resistance, U stands for voltage and w_r stands for electrical angular speed (the electrical angular speed is equal to the mechanical angular speed in the per-unit system).

The field voltage is described as

$$E_f = L_{md} \frac{U_f}{R_f} \tag{8.19}$$

The rotor motion equations which describe the variation of electromagnetic torque, mechanical angular speed and load angle are given as follows:

$$\begin{cases} T_e = \psi_d i_q - \psi_q i_d \\ t_b \dfrac{dw_m}{dt} = \dfrac{1}{2H}(T_m - T_e) \\ \dfrac{d\delta}{dt} = w_m - 1 \end{cases} \tag{8.20}$$

where T_e stands for the electromagnetic torque, w_m stands for the mechanical angular speed, δ stands for the rotor angle deviation and H stands for the inertia time constant in per-unit system which can be described as

$$H = \frac{J w_{m,b}}{2 t_b S_b} \tag{8.21}$$

where J is the inertia, and $w_{m,b}$ is the base value of mechanical angular speed which can be described as

$$w_{m,B} = \frac{1}{t_b p} \tag{8.22}$$

where p stands for the number of pole pairs.

The three-phase synchronous machine is an AC electric machine in which the armature windings relate to the three-phase source and the DC source is needed for the rotor to produce a steady magnetic field.

8.2.4   Gas reservoir

The gas reservoir is the place to store compressed air. It can be an air tank or salt dome/bed carven in most cases.

Compare to other components, the model of the gas reservoir is relatively simple. The model is based on ideal gas theory. Literature [18] has illustrated how to build it. A model of two equations of mass balance and energy balance based on real gas was proposed in [25].

To model a more realistic gas reservoir like salt dome/bed carven, the heat transfer between the stored compressed air and the salt cavern wall needs to be considered.

8.3   Data-driven modelling

Using the derived models that are provided in the presented work, CAES system simulation can be achieved. However, the simulation can be computationally expensive and therefore potentially very time-consuming. To overcome this problem, a data-driven approach based on neural networks (NNs) is implemented.

Unlike models based on differential equations (lumped-parameter model, thermodynamic model, etc.), DDMs do not require the MATLAB Simulink® to call its solvers; thus, the computational cost drops sharply. Therefore, this kind of model can be applied for accelerating whole-system speed of simulation of large a CAES system and subsystems.

The main restriction for a DDM is that they cannot handle an input if it is outside the range of the model's training set. Therefore, the training set is essential for a DDM to be applied properly. Another restriction is that the length of time step in simulation is decided by the training data set. Therefore, the DDM can take less benefit from variable-step solver if its maximum time step is large.

To take advantage of other verified models mentioned above, the training set of a DDM is generated from them. In this way, the training set is designed to represent target operations and generated from a single-component model for training the DDM, and the DDM can be used in system simulation afterwards.

In this simulation package, the NN is utilized for the data-driven approach. A rigorous approach is required for building an NN model of the target component to ensure that the outcome is an accurate representation of the original model. There are four steps in this process:

1. Evaluate the value of building an NN model for a component.
2. Build a training system.
3. Train the NN model for the component.
4. Validate the NN model.

The scheme of the basic process is presented in Figure 8.4.

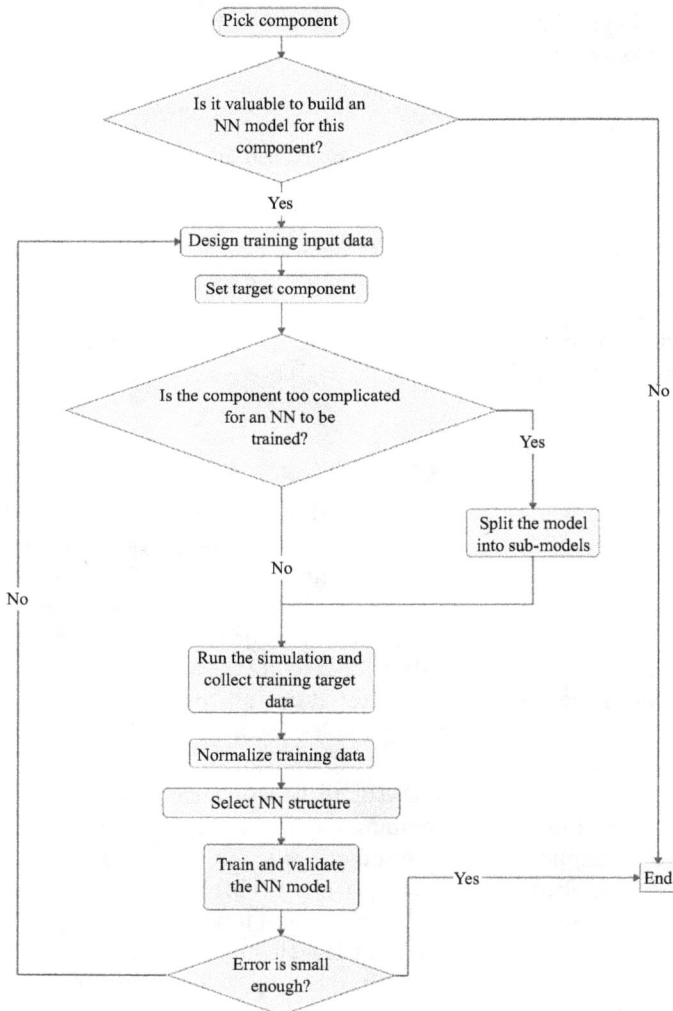

Figure 8.4   The basic process for building an NN model for the target component

8.3.1   Evaluating the value of building an NN model for a component

The purpose of building an NN model for a component is to accelerate the simulation speed while maintaining the accuracy of the model for achieving real-time simulation and HIL (hardware in the loop) prototyping. As the fixed-step simulation is required for such applications, the trained NN for a component must speed up the model simulation under a fixed-step solver, otherwise no benefit is achieved. To achieve this, computational consumption needs to be considered.

By comparing computational consumption, the time complexity is introduced. In computer science, this describes the amount of time it takes to run an algorithm. In this project, the time complexity is used to evaluate the simulation speed of FPMs and DDMs.

For a trained NN, the calculation complexity is equal to a matrix multiplication calculation. Its algorithm complexity is $O(n)$ for each step, where n is the number of neurons of the NN. The complexity is $O(nT)$ for the whole simulation, where T is the number of steps it takes in this simulation. For an FPM of a component in the CAES system, the algorithm complexity depends on its characteristics.

Most FPMs for electrical machines, compressors, air tanks and other components in the CAES system are based on differential equations and implicit nonlinear algebraic equations. These models require a simulation environment to solve the equations at each step. The time complexity of the FPM is defined in equation

$$\text{Time complexity} = O((M_0 + \sum_{i=1}^{n} I_i M_i) I_0 ST) \qquad (8.23)$$

where M_0 is the number of calculation operator that is outside an algebraic loop, I_0 is the average iteration number for solving the optimization problem for the model, M_i ($i \geq 1$) is the number of calculation operator that is inside the ith algebraic loop, i is the average iteration number for solving the ith algebraic loop in each step, n is the number of an algebraic loop in this FPM, S is the number of calculation points for the solver to calculate the integration result of one step (i.e. $S = 3$ for ode3, $S = 5$ for ode5) and T is the number of steps for one simulation.

For FPMs of components in the CAES system, the algebraic loop is usually inevitable, because laws like energy balance and mass balance will create algebraic loops in a model, and such laws are usually fundamental in FPMs of components in CAES systems. NN models prevent algebraic loops and thus shorten the simulation time.

Some FPMs use numerical methods that require iteration calculation to meet the limiting factor in every step. For instance, the radial turbine model presented in [26] requires the calculation of the value of an intermediate pressure in each step to satisfy the mass balance of the turbine. The time complexity for models like this needs to be calculated, respectively, as it may have higher complexity than other FPMs.

For the NN model, the time complexity is decided by the number of neurons. For each neuron, the output equals the input multiply by weight and then adds by bias and then passes through the activation function. The time complexity is approximately $O(5nT)$, where n is the number of neurons. For FPMs which need higher-level solver to resolve, using NN model can save more time as they don't need to call the solver.

Comparison of the time complexity between FPM and NN models can give a clear indication of the effect that an NN model can have on accelerating the simulation. The more complex the system is, the greater the time reduction is achieved by using NN models instead of the FPMs, as the time complexity will be multiplied to each other when components' FPMs are integrated.

8.3.2   Designing a training system

If building an NN model is valuable, the next step is to design the training system for it. In this project, the training system consists of several training cells. A training cell is defined as a system that can train an NN model. A general structure of a training cell is depicted in Figure 8.5.

Figure 8.5 shows the logical structure of the general training system. In real situations, several training cells can share the same training data collecting and normalizing block, or a training cell can use data generated from several different systems.

8.3.3   Training the NN

To train the model, a supervised learning method is chosen for this project. The point is to update variable parameters in the DDM to minimize the error between the DDM output and the normalized output. Supervised training helps the NN to learn a function that maps output from an input based on example inputs/outputs

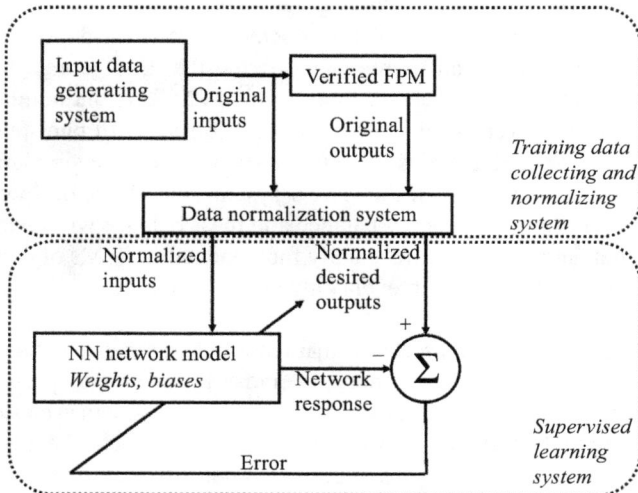

Figure 8.5   A general structure for a training cell

pairs [27]. It is natural to choose a supervised training method for this project, as all components in CAES have inputs/outputs pairs.

The supervised learning method for NNs can be separated into two parts: the optimization algorithm, and the weights and biases updating algorithm. The optimization algorithm calculates recurrently to give directions of weights and biases update for reducing the NN output error, while the weights and biases updating algorithm implement optimization result onto NN's parameters.

Many algorithms can be chosen as the optimization algorithm to train the NN, such as the Newton and quasi-Newton methods, and stochastic gradient descent (SGD) series methods. Newton and quasi-Newton methods can get the global optimal solution but will cost more computation resource and time, while SGD series method can get a solution faster, while sometimes it may be a suboptimal one. More details about optimization algorithm can be found in [28].

For the weight and biases updating algorithm, the backpropagation method is still the most effective method at present. Literature [29] gives a detailed explanation of this method.

8.3.4 Validating the NN model

After the NN is trained, it must then be validated. There are three steps to validate a trained NN model. Figure 8.6 shows how the validation is processed.

1. Validate the NN data with the training data set in Simulink®. This test can ensure that the NN is trained as designed. If the NN cannot represent the

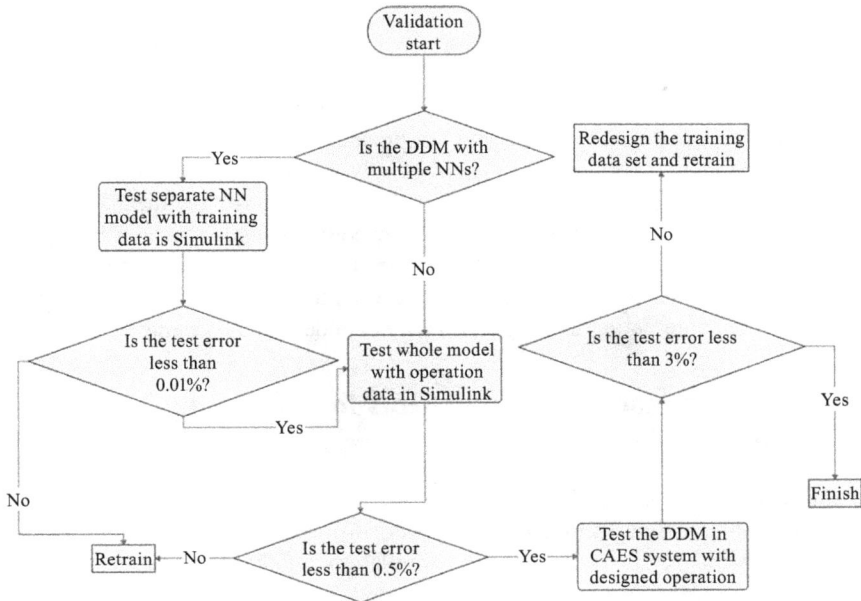

Figure 8.6 DDM validation process

sub-model correctly, there are two sets need to be checked: (i) the Simulink solver needs to be either in variable-step mode or fixed-step mode with the same step length of the training data. Otherwise, the NN will not represent the correct response of the target component. (ii) The initial condition of the NN model needs to be checked. After adjustment, if the NN is still not working correctly, it is because the training of this NN is not good enough. To avoid overfitting, the training initial condition is randomly selected each time, so it is possible for some training set, the initial condition is not good for convergence. In this case, the NN needs to be retrained. In this step, if the NN model has an error of less than 0.01%, then we proceed to the next step.

2.   For those components which have been split into sub-models, the next step is to test the whole model after assembling all sub-models. In this step, operation data for the component can be used to test. If the whole model does not work well, the range of any sub-model's input and output needs to be checked to be within the range of the training data set. Because of errors, sometimes the system comprised of NNs may have unwanted outputs which will not happen in a single sub-model test, because of the correlation among each sub-model. If all inputs and outputs for all sub-models are within the designed range, but the whole model is not working, this means the designed training input data did not represent the operation condition well. In both cases, the input training data need to be redesigned.

For components that are represented with one NN, these are also tested by using operation data. If the NN model is not working, this means the designed training input data did not represent the operation condition well and need to be redesigned.

In this step, if the NN model has an error of less than 0.5%, it can proceed to the next step.

3.   If the NN passed the two tests above, final validation is conducted by putting the new model into a CAES system and testing under the operating conditions. If the NN model is not working, this means the whole system is not stable with errors arising within the model. If the NN itself becomes unstable, then the training input data needs to be redesigned. If other components in this system become unstable, some adjustments need to be made on the specific component. If the NN model has an error of less than 3%, the validation completes, and the NN is capable of representing the target component under certain operating conditions.

8.4   Choosing models of components for simulating a CAES system

As introduced above, the thermodynamic model, lumped-parameter model and DDM are implemented. In this way, the recent work allows using different model types for CAES system and subsystem simulation for different purposes. In this section, five types of models will be compared further to guide for choosing a proper model type for simulation.

The purpose of the CAES system simulation is focused on evaluating and improving existing system performance, validating the CAES system integration with renewable energy sources, and designing a new CAES system. There are two main categories that can be recognized from literature in this area. One is focused on a key component, such as compressor or expander, to study how the operation of the component can impact on the whole-system performance, and how to optimize it. The works in references [19,30] and [31] are examples in this type. The other one is focused on how the whole CAES system performs as a whole under different configurations and working conditions, and how to improve its efficiency, flexibility and robustness. The works in references [6,20] and [23] are examples in this type.

For dynamic analysis of a key component, the dynamic behaviour of other components in the system can be ignored selectively. It is because the timescale difference among components with the different physical process is significantly different. For generator, motor and the power grid, the time constant is milliseconds. For the compressor and turbine, the time constant is a few seconds to minutes. For TESs and HEXs, the time constant can be much larger. Therefore, when the simulation is focused on electrical components, the dynamic behaviour of thermal components can be ignored, vice versa.

The type of model should be chosen based on the purpose and perspectives of the system simulation. The good match between the model type and the simulation goal can lead to informative simulation results with optimized simulation speed.

Figure 8.7 reveals the system description ability and computational cost requirement of different kinds of models mentioned above.

There are four quadrants in Figure 8.7. Phase 1 represents models which have high system description and low computational cost. Models in this phase are suitable for both system modelling and subsystem modelling, as they provide the most detailed information with a small computational cost. Lumped-parameter model for relatively simple component (air tank for example) can be regarded as in this phase.

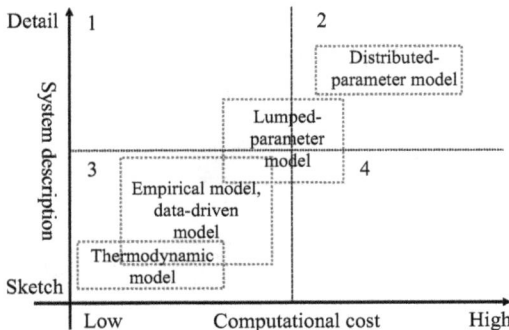

Figure 8.7   System description ability and computational cost for different types of models

Phase 2 represents models which have high system description ability with a high requirement in computational cost. Distributed-parameter model is typically in that phase. Some lumped-parameter model for a complex component like compressors or motors can also belong to this phase. This type of model is mainly used for the analysis and optimization of a well-designed component, as they typically require the most detailed system description.

Phase 3 represents models which have low system description ability with a low requirement in computational cost. Thermodynamic model, empirical model and DDM can be regarded in this phase. Models in this phase are mainly used for system-level simulation and optimization, as an affordable computational cost needs to ensure its implementation.

Phase 4 represents models which have low system description ability yet require for high computational cost. Models in this phase may not be suitable for a CAES system model, as they are not fast enough to be used in system simulation, yet not detailed enough to be applied in component simulation. For instance, a lumped-parameter model for a HEX with too many subsections will not give more system detail than the distributed-parameter models but can consume significant calculation resources. Researchers should avoid generating model in that phase.

Based on the properties of different models, a strategy tree for choosing a suitable type of model is shown in Figure 8.8.

The decision tree explains how to choose a suitable model type for different simulation purposes. The end of each arrow is one or two types of simulation models. At the bottom of the figure from left to right, the model's level of the system description reduces, while the computational cost reduces as well.

Table 8.2 concludes the suggested application for different types of models presented in this work. It will help to understand the decision tree in more details.

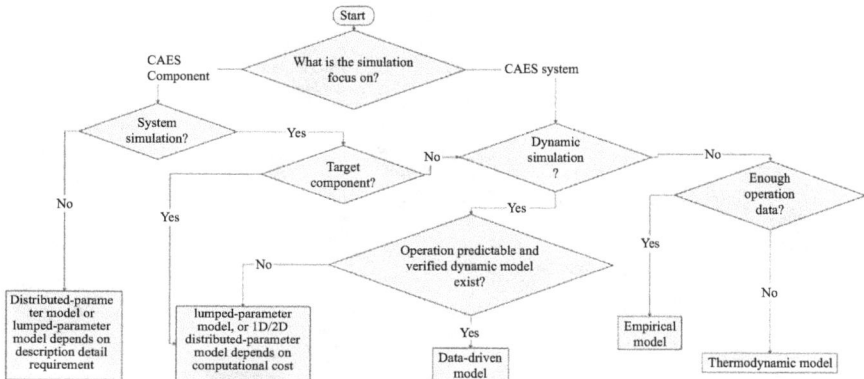

Figure 8.8   Strategy tree for choosing a suitable type of model

Table 8.2 Suggested application for different types of models

	Thermodynamic model	Empirical model	DDM	Lumped-parameter model	Distributed-parameter model
Steady-state system simulation	Yes	Yes	No	No	No
Dynamic system simulation focused on a specific component	Yes	No	Yes for surrounding component	Yes for the specific component	Yes for the specific component (1D/2D)
Whole-system dynamic simulation	No	No	Yes	Yes	Yes (1D/2D)
Dynamic component simulation	No	No	No	Yes	Yes

8.5 CAES system modelling case for applying a DDM

The works in references [17] and [18] give some examples of how to use thermodynamic models and lumped-parameter models to implement CAES system simulation. This section will give an example of how to use the DDM to approach CAES system simulation.

As mentioned above, lumped-parameter models are suitable for building a dynamic model for the CAES system. DDM, on the other hand, can accelerate the simulation speed of the dynamic model, which makes it possible to make real-time simulation in the future. In this section, a kW-scale adiabatic CAES system discharging process dynamic model is built, and the DDM is used on the three-stage turbine in the system to accelerate the simulation speed. The system scheme is shown in Figure 8.9. The system is based on the 500-kW non-supplementary fired CAES demonstration system (TICC-500). This system is the world's first multistage regenerative CAES system for power generation which was built in 2014 [32].

The discharging part of the system consists of an air tank, a hot water tank type TES, five-stage inter-stage HEXs, a three-stage radial turbine and a generator. The system is simulated by FPMs first, and then the three-stage radial turbine model is replaced by DDMs. The parameter for the three-stage turbine is shown in Table 8.3. The input air will first pass through HEXs before going into the turbine; thus, the temperature of inlet compressed air is equal to the outlet of HEXs.

According to Section 8.3, the three-stage turbine is a good choice for DDM replacement in the TICC-500 system. First, the time complexity of the one-stage

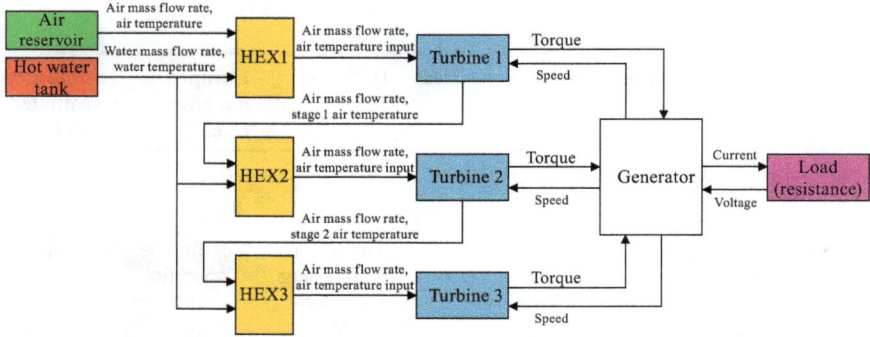

Figure 8.9   Adiabatic CAES system scheme [32]

Table 8.3   Parameters for the three-stage turbine in TICC-500

Operational parameter	Value	Geometric parameter	Value
First stage turbine			
Inlet pressure, MPa	2.30	Rotor inlet diameter, m	0.1728
Rotor inlet angle, degree	90	Rotor outlet mean diameter, m	0.0773
Stator frictional coefficient	0.88	Rotor inlet width, m	0.0044
Rotor frictional coefficient	0.88	Rotor outlet width, m	0.0148
Lower limit of pressure ratio	1	Rotor outlet angle, degree	$90 - 56 = 34$
Upper limit of pressure ratio	3	Nozzle outlet angle, degree	$90 - 77 = 13$
Second stage turbine			
Inlet pressure, MPa	1.0	Rotor inlet diameter, m	0.2226
Rotor inlet angle, degree	90	Rotor outlet mean diameter, m	0.1050
Stator frictional coefficient	0.9	Rotor inlet width, m	0.0074
Rotor frictional coefficient	0.81	Rotor outlet width, m	0.0253
Lower limit of pressure ratio	1	Rotor outlet angle, degree	34
Upper limit of pressure ratio	4	Nozzle outlet angle, degree	13
Third stage turbine			
Inlet pressure, MPa	0.5	Rotor inlet diameter, m	0.2515
Rotor inlet angle, degree	90	Rotor outlet mean diameter, m	0.1144
Stator frictional coefficient	0.86	Rotor inlet width, m	0.0192
Rotor frictional coefficient	0.86	Rotor outlet width, m	0.0642
Lower limit of pressure ratio	1	Rotor outlet angle, degree	34
Upper limit of pressure ratio	5	Nozzle outlet angle, degree	13

turbine can be calculated based on the model equations in Section 8.2.1 and Equation (8.23). For the one-stage turbine model, $M_0 = 82, I_0 = 8$, so the time complexity is O(656ST). For a DDM with 20 neurons, the time complexity is O (100ST); thus, a DDM for three-stage turbine can accelerate the simulation speed for 6.56 times theoretically for the turbine alone.

Figure 8.10 shows the simulation result of the TICC-500 system discharging process. From Figure 8.10, the system starts up within 5 sand then achieves a stable speed and power output. The air pressure in the tank drops at a steady rate.

Figure 8.11 illustrates the comparison between FPM and DDM of the dynamic response from a different stage of the turbine. In this simulation, the water flow of the HEXs is different, with 5.4 kg/s for Stage 1 and 1 kg/s for Stages 2 and 3.

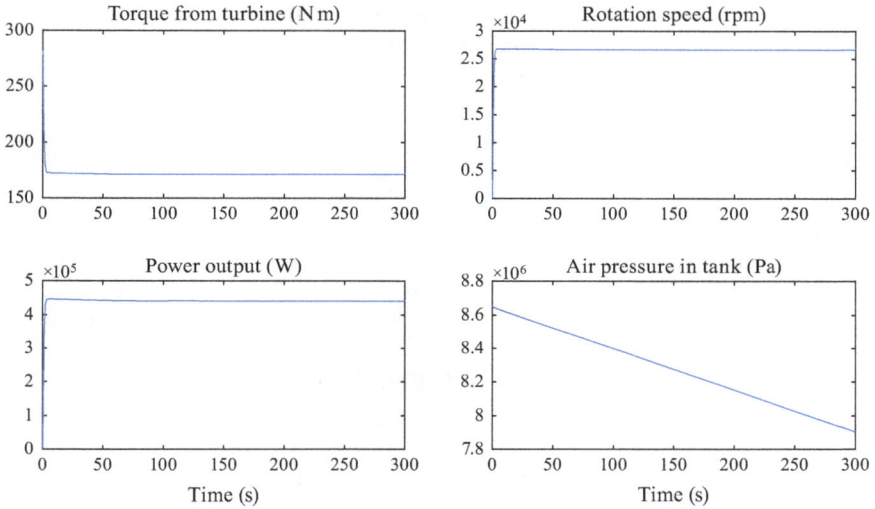

Figure 8.10   TICC-500 discharging process simulation result

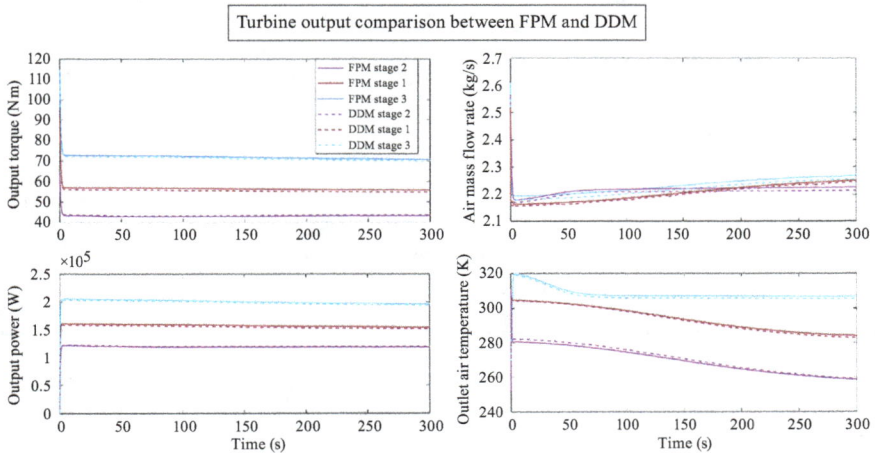

Figure 8.11   Turbine output comparison between FPM and DDM

The three-stage turbine is first simulated with the FPM, and then replaced by a DDM, and the output of the DDM mixed system simulation is compared with the prior model. Its error is shown in Table 8.4.

The test shows that the multi-stage turbine NN model can replace the part of the turbine in the original model accurately.

To evaluate how much the simulation speed is improved, first the time consumption of the turbine in the TICC-500 system is tested. Using the simulation profile function, the MATLAB Simulink time of each component in the TICC-500 system can be distinguished in Table 8.5. The three-stage turbine has consumed approximately 22.8% of simulation time under fix-step solver, which is the second most consuming component.

The time different under ode8 solver between the FPM model and the DDM-FPM combined model for the TICC-500 system is shown in Table 8.6. From this table, the DDM of the three-stage turbine can improve the TICC-500 system simulation speed for approximately 20% on average. The computation time for other components have no significant difference, and the three-stage turbine computation speed has been accelerated by 6.47 times, which is close to the theoretical value.

Table 8.4   The error between FPM and DDM for three-stage turbine

	Turbine stage 1 Torque(N m)	Air mass flow rate(kg m/s)	Isentropic efficiency	Power(W)	Air temperature (K)
MSE	0.1799	0.0019	0.0025	477	0.7144
MSE%	0.40%	0.09%	0.39%	0.45%	0.22%
	Turbine stage 2				
MSE	0.3957	0.0061	0.0055	795	1.1580
MSE%	1.15%	0.34%	0.86%	1.02%	0.36%
	Turbine stage 3				
MSE	0.2913	0.0044	0.0031	1,763	0.8605
MSE%	0.27%	0.16%	0.46%	0.69%	0.41%

MSE, mean square error.

Table 8.5   Computation cost percentage for different component in TICC-500 model

Component name	Percentage of computation cost
HEX	65.6%
Radial turbine	22.7%
Air tank	1.4%
DC generator	0.1%

Table 8.6 Computation time comparison between FPM and DDM

Simulation time (s)	10	20	30	40	50
Computation time with DDM (s)	857.5	687.3	513.7	357.7	177.7
Computation time without DDM (s)	1,054.7	845.0	644.9	474.4	210.0
Improvement (%)	18.7	18.6	20.3	24.6	15.4

8.6 Conclusion

The chapter introduced recent work from the University of Warwick on dynamic modelling of the grid-scale CAES system and its key components. Both key components modelling methods and CAES system dynamic simulation has been described. A novel data-driven approach is developed to accelerate the CAES system dynamic modelling. Guidance for how to choose a proper simulation model type for a different purpose is given, and a test for accelerating simulation speed by using the DDM has been presented. From the test using DDM to replace an axial turbine FPM in the TICC-500 system model, the simulation speed of the system can be accelerated by average 20%. Utilizing DDM to replace other components in CAES system can further improve the simulation speed of the complex system, and may achieve real-time simulation in the future.

References

[1] F. R. Kalhammer and T. R. Schneider, 'Energy storage', *Annu. Rev. Energy*, vol. 1, no. 1, pp. 311–343, 1976.

[2] W. Mattick, H. G. Haddenhorst, O. Weber and Z. S. Stys, 'Huntorf: the world's first 290-MW gas turbine air-storage peaking plant. [Compressed air pumped hydro storage]', *Proc. Am. Power Conf.; (United States)*, vol. 37, 1975.

[3] S. Herrmann, H-J. Kretzschmar and D. P. Gatley, 'Thermodynamic properties of real moist air, dry air, steam, water, and ice (RP-1485)', *HVAC R Res.*, vol. 15, no. 5, pp. 961–986, 2009.

[4] S. Herrmann, H-J. Kretzschmar, V. Teske *et al.*, 'Properties of humid air for calculating power cycles', *J. Eng. Gas Turbines Power*, vol. 132, no. 9, 2010.

[5] X. Ji and J. Yan, 'Thermodynamic properties for humid gases from 298 to 573 K and up to 200 bar', *Appl. Therm. Eng.*, vol. 26, no. 2–3, pp. 251–258, 2006.

[6] H. Peng, Y. Yang, R. Li and X. Ling, 'Thermodynamic analysis of an improved adiabatic compressed air energy storage system', *Appl. Energy*, vol. 183, pp. 1361–1373, 2016.

[7] H. Safaei and M. J. Aziz, 'Thermodynamic analysis of three compressed air energy storage systems: Conventional, adiabatic, and hydrogen-fueled', *Energies*, vol. 10, no. 7, p. 1020, 2017.

[8]   S. Houssainy, M. Janbozorgi, P. Ip and P. Kavehpour, 'Thermodynamic analysis of a high temperature hybrid compressed air energy storage (HTH-CAES) system', *Renew. Energy*, vol. 115, pp. 1043–1054, 2018.

[9]   Y. Zhang, K. Yang, X. Li and J. Xu, 'The thermodynamic effect of thermal energy storage on compressed air energy storage system', *Renew. Energy*, vol. 50, pp. 227–235, 2013.

[10]   R. Kushnir, A. Ullmann, and A. Dayan, 'Thermodynamic and hydrodynamic response of compressed air energy storage reservoirs: a review', *Reviews in Chemical Engineering*, vol. 28, no. 2-3, pp. 123–148.

[11]   W. He and J. Wang, 'Optimal selection of air expansion machine in compressed air energy storage: A review', *Renewable and Sustainable Energy Reviews*, vol. 87. Elsevier Ltd, pp. 77–95.

[12]   H. van Putten and P. Colonna, 'Dynamic modeling of steam power cycles: Part II – Simulation of a small simple Rankine cycle system', *Appl. Therm. Eng.*, 2007. vol. 27, no. 14–15, pp. 2566–2582, 2007.

[13]   E. Baskharone, *Principles of Turbomachinery in Air-breathing Engines*. Cambridge: Cambridge University Press; 2006.

[14]   S. L. Dixon, *Fluid Mechanics, Thermodynamics of Turbomachinery*. Burlington, MA: Butterworth-Heinemann; 2010.

[15]   F. J. Wallace, 'Theoretical assessment of the performance characteristics of inward radial flow turbines', *Proc. Inst. Mech. Eng.*, vol. 172, no. 1, pp. 931–952, 1958.

[16]   H. Safaei and D. W. Keith, 'Compressed air energy storage with waste heat export: An Alberta case study', *Energy Convers. Manag.*, vol. 78, pp. 114–124, 2014.

[17]   X. Luo, J. Wang, C. Krupke, *et al.*, 'Modelling study, efficiency analysis and optimisation of large-scale adiabatic compressed air energy storage systems with low-temperature thermal storage', *Appl. Energy*, vol. 162, pp. 589–600, 2016.

[18]   X. Luo, M. Dooner, W. He, *et al.*, 'Feasibility study of a simulation software tool development for dynamic modelling and transient control of adiabatic compressed air energy storage with its electrical power system applications', *Appl. Energy*, vol. 228, pp. 1198–1219, 2018.

[19]   T. A. C. Maia, O. A. Faria, J. E. M. Barros, M. P. Porto and B. J. Cardoso Filho, 'Test and simulation of an electric generator driven by a micro-turbine', *Electr. Power Syst. Res.*, vol. 147, pp. 224–232, 2017.

[20]   M. Martínez, M. G. Molina and P. E. Mercado, 'Dynamic performance of compressed air energy storage (CAES) plant for applications in power systems', in *2010 IEEE/PES Transmission and Distribution Conference and Exposition: Latin America*, T and D-LA 2010, 2011, pp. 496–503.

[21]   P. Omsin, S. M. Sharkh and M. Moshrefi-Torbati, 'A hybrid SS-CAES system with a battery', in *Proceedings of the 16th International Conference on Electrical Engineering/Electronics, Computer, Telecommunications and Information Technology, ECTI-CON 2019*, 2019, pp. 159–162.

[22] V. Kokaew, S. M. Sharkh and M. Moshrefi-Torbati, 'A hybrid method for maximum power tracking of a small scale CAES system', in *2014 9th International Symposium on Communication Systems, Networks and Digital Signal Processing, CSNDSP 2014*, 2014, pp. 61–66.

[23] Y. Mazloum, H. Sayah and M. Nemer, 'Dynamic modeling and simulation of an isobaric adiabatic compressed air energy storage (IA-CAES) system', *J. Energy Storage*, vol. 11, pp. 178–190, 2017.

[24] S. Collins, 'Commercial options for energy storage multiply', *Power*, vol. 137, pp. 55–57, 2019.

[25] Y. Zhang, Y. Xu, X. Zhou, H. Guo, X. Zhang and H. Chen, 'Compressed air energy storage system with variable configuration for accommodating large-amplitude wind power fluctuation', *Appl. Energy*, vol. 239, pp. 957–968, 2019.

[26] W. He, J. Wang and Y. Ding, 'New radial turbine dynamic modelling in a low-temperature adiabatic compressed air energy storage system discharging process', Energy Convers. Manag., 2017.

[27] S. Russell and P. Norvig, *Artificial Intelligence: A Modern Approach*, Third Edition. Englewood Cliffs, NJ: Prentice Hall; 2010.

[28] S. Ruder, 'An overview of gradient descent optimization algorithms', arXiv Prepr. arXiv1609.04747, 2016.

[29] P. J. Werbos, 'Backpropagation through time: What it does and how to do it,' *Proc. IEEE*, vol. 78, no. 10, pp. 1550–1560, 1990.

[30] V. Tola, V. Meloni, F. Spadaccini and G. Cau, 'Performance assessment of adiabatic compressed air energy storage (A-CAES) power plants integrated with packed-bed thermocline storage systems', *Energy Convers. Manag.*, vol. 151, pp. 343–356, 2017.

[31] K. Yang, Y. Zhang, X. Li and J. Xu, 'Theoretical evaluation on the impact of heat exchanger in Advanced Adiabatic Compressed Air Energy Storage system', *Energy Convers. Manag.*, vol. 86, pp. 1031–1044, 2014.

[32] S. Mei, J. Wang, F. Tian, *et al.*, 'Design and engineering implementation of non-supplementary fired compressed air energy storage system: TICC-500', *Sci. China Technol. Sci.*, vol. 58, no. 4, pp. 600–611, 2015.

Chapter 9

Application of compressed air energy storage systems in a day-ahead dispatch schedule under demand response and renewable obligation

Thabo G. Hlalele[1], Jiangfeng Zhang[2], Raj Naidoo[1] and Ramesh C. Bansal[3]

In this chapter, a combined day-ahead dispatch schedule for compressed air energy storage (CAES) systems with renewable energy sources (RESs) under demand response and renewable obligation is presented. The model uses CAES systems to overcome the uncertainty related to wind and photovoltaic (PV) energy systems. The proposed model is applied to a renewable obligation policy which ensures that a certain percentage of the energy generated comes from RESs. In order to participate in the day-ahead market, CAES systems are integrated with RES generators, and demand response is used for deferring flexible demand from peak electricity price period to low-price periods. The proposed model uses real data from a large-scale demand response programme which allows the system operator to directly control the participation of electric water heaters from the substation level. The effectiveness of the proposed model is tested on a modified IEEE 30-bus system and the results show that the CAES systems can help in increasing the RES penetration and improve profitability.

Keywords: Compressed air energy storage; demand response; day-ahead dispatch; renewable energy sources

9.1 Introduction

There is a shift from passive operation of the electric power system to a more active power distribution network. This is mainly due to an increase level of distributed

[1]Department of Electrical, Electronic and Computer Engineering, University of Pretoria, Pretoria, South Africa
[2]Department of Automotive Engineering, Clemson University, Greenville, SC, USA
[3]Department of Electrical Engineering, University of Sharjah, Sharjah, United Arab Emirates

generators, responsive loads and energy storage systems integrated with RESs in the power system. The inherent nature of the intermittent energy sources such as wind and PV makes it very challenging to integrate them into the day-ahead energy market in a deterministic manner as they cannot compete with conventional thermal generators. Therefore, it is essential to actively manage the integration of stochastic energy resources with energy storage systems such as CAES in order to allow them to participate in the day-ahead market.

In recent years, there has been a significant attention on the usage of CAES system as a better alternative to other energy storage systems such as lithium-ion batteries due to its cost-effectiveness and technological maturity. The reason for the increased usage of CAES system comes from the fact that it can provide bulk-scale energy storage and a low storage cost per unit energy [1]. More importantly, the coordination of the intermittent energy sources with energy storage can help mitigate the volatility associated with the stochastic energy sources and can take advantage of the real-time price (RTP) of electricity by charging during periods of low energy cost and discharge during periods of high energy cost.

9.1.1 Literature review

The integration of energy storage specifically CAES has shown that the coordination of RES and CAES can reduce the level of curtailment by storing excess energy for later usage. For example, in [2], they show wind curtailment can be reduced substantially by more than 20% by coordination of the wind power plant with CAES. The authors further show that this type of coordination can increase the profitability of the hybrid power plant and improve long-term growth of RES development. The coordination of RES and storage has been studied in recent works. A self-scheduling adaptive robust approach is presented in [3], in which the coordination of wind and CAES is used in the offering and bidding strategy. The uncertainty related to wind power and price forecast is modelled using adaptive robust optimisation and the flexibility of the proposed method uses a budget of uncertainty to limit the overall output deviation for both wind energy production and electricity price forecast errors. In [3], the authors show that integrating and coordinating wind power production with energy storage can realise the best performance in terms of shifting power production from low-price to high-price period, compensating for over and under bid of wind farms and using storage to participate in the market.

A bidding and offering strategy is presented in [4] for a merchant energy storage facility that uses CAES for participating in the market. The authors show that the electricity price forecast errors can be minimised by using information gap decision theory (IGDT) in which the risk associated with bidding and offering is offset by minimising the electricity price forecast errors. The price uncertainty is modelled using IGDT and hourly bidding and offering curves are presented for market participation of the CAES system. In [5], a CAES model is presented for a look-ahead optimisation model for participation in the day-ahead and real-time energy markets for ancillary services and reserve markets. The proposed model extends the scheduling day-ahead market and maximises the utilisation of the CAES in market participation. A day-ahead market participation equipped with

CAES is proposed in [6]. The authors propose the usage of CAES in the active distribution network that participates in the wholesale market where they optimally coordinate CAES with electric vehicle charging stations to maximise their profit.

A risk-constrained bidding strategy is presented in [7], in which a multi-stage stochastic optimisation model is presented for a joint operation of wind farm and CAES for participation in three energy markets, namely, day-ahead, intra-day and balancing market. The inherent risk associated with the uncertainty in electricity price and intermittent resource is modelled using stochastic programming. Moreover, the paper uses conditional value-at-risk to control financial risks associated with the three energy markets. The authors show that the proposed model can achieve joint cost-effective operation of wind power plant and CAES to maximise its profit.

Although a multitude of research has been conducted in the joint operation of intermittent energy sources and CAES or the operation of CAES for the ancillary and reserve market, the technical literature lacks a model that focuses on the combined integration of uncertain energy source and energy storage under renewable obligation framework. The renewable obligation framework ensures that a minimum amount of dispatched energy comes from wind and solar energy and ensures that at all times a certain renewable obligation is attained. In case this obligation is not attained, a penalty is imposed by the system operator to ensure that adequate energy mix is achieved all the time. Although renewable obligation can attain high RES penetration, it is prone to the intermittent nature of the RESs, it is therefore, important to minimise the volatility associated with RES by jointly operating RES with storage such as CAES to improve the profitability of renewable energy aggregator. This chapter extends from [8] and [9] by integrating CAES into the renewable obligation framework so that the renewable energy producers can maximise their profit when participating in the day-ahead market. The benefit of introducing CAES to the renewable obligation framework is to participate in the electricity market and take advantage of the arbitrage profits arising from market conditions.

9.1.2   Research objectives

To address the aforementioned challenges, this chapter aims to develop a joint coordination of stochastic energy resources and CAES system in order to maximise the profit of the hybrid power plant. Moreover, the inclusion of the renewable obligation framework ensures that renewable energy penetration is prioritised while the CAES minimises the volatility that comes from wind power plants and stores excess energy for usage in other time period. The inherent dependency of RESs is minimised by providing storage capacity to ensure that the hybrid power plant is dispatchable. The main contributions are summarised as follows:

1.   A unit commitment model is presented for a joint coordination for intermittent energy sources and CAES for a day-ahead market participation. A renewable obligation framework is used for ensuring that an adequate energy mix is achieved over the day-ahead dispatch period.

2.   CAES system is used to overcome the intermittency of PV and wind power plant and coordinate the power dispatch to maximise economic profit of the facility.

9.1.3   Chapter organisation

This chapter is organised as follows. In Section 9.2, we provide a detailed background on a CAES system and its integration to power system operation. In Section 9.3, we introduce a simplified mathematical model for a joint CAES renewable obligation model for a hybrid power plant. These models are further used to build a tool for the operation of the hybrid power plant for a day-ahead market participation. In Section 9.4, the proposed model is investigated using a standard IEEE 30-bus system to show its effectiveness. Finally in Section 9.5, the conclusions are drawn.

9.2   Compressed air energy storage

The original intention for developing CAES system was to leverage from the inexpensive and off-peak cost of electricity and transfer it to peak load periods. This was supported by the excellent capability to start-up and ability to support with ancillary services. It is one of the excellent tools for load shifting applications and can support the intermittent energy sources. A CAES system typically consists of three main processes, namely the charging process, storage and discharging process. In the charging process, excess electricity from the RESs or during low tariff period is used to drive the electric motor which is coupled to a compressor system.

The compressor system draws ambient air and compresses it through a multi-stage process which depends on the system requirement. As a result, heat is produced as one of the by-products, which can either be discarded or stored for later usage in the discharging process. The compressed air is stored in either salt caves, mines, wells, gas chambers, tanks or reservoirs [10]. The discharge process works in reverse to the charging process. First, air is heated if it is not stored as part of the storage process. The heated air is used to drive the expander and is further changed into electrical energy for usage and export to the power system grid. Typically, two types of air heating systems are used in the CAES process, namely an external heating source and a heat recovery process. The diabatic CAES (D-CAES) process uses an external heat source from fossil fuels while the adiabatic CAES (A-CAES) uses heat recovered from the charging-up process. Figure 9.1 shows a typical A-CAES process which integrates the intermittent energy sources in the charging phase to smooth the RES generators and store excess energy.

The CAES system is typically classified as either small (<100 MW) or large scale (>100 MW). Another classification of heat transfer is also called isothermal-CAES (I-CAES), which is an extension of A-CAES as it reduces heat energy losses by cooling air in the compression phase in order to prevent excessive temperature rise, and using the recycled heat during compression to maintain constant temperature expansion [11]. The I-CAES process is based on piston machinery and utilises liquid piston for the use of additional heat transfer. The typical efficiency for D-CAES is 50% while A-CAES efficiency ranges between 60% and 70% and

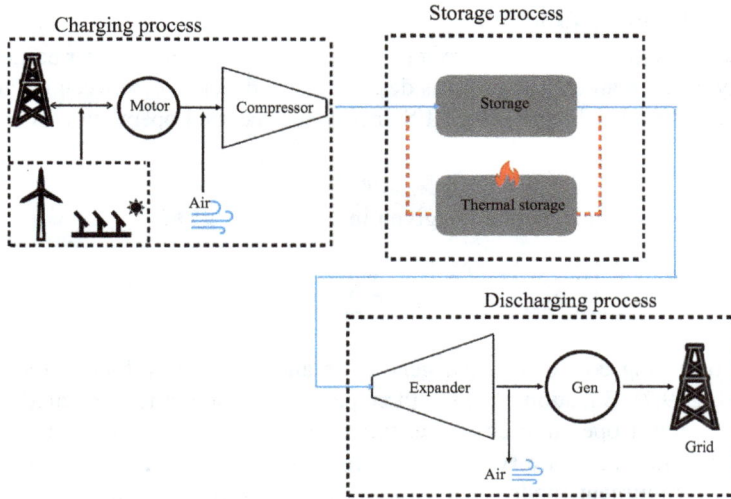

Figure 9.1   A-CAES system using RES intermittent energy sources for the charging process

I-CAES efficiency can reach up to 80% [10]. The main advantage of a CAES system is its ability to provide cold, heat and electricity all in one, which makes its application even more vast in distribution energy systems.

9.3   Problem formulation

In this section, the formulation of the joint RES and CAES system is presented. The power system consists of thermal generators, PV, wind and CAES systems. The main objective is to maximise the profit of the joint RES and CAES system in a day-ahead dispatch schedule. The RES–CAES aggregator participates in the bidding and offering process in the day-ahead market. Note that when RES generates excess energy in the early hours of the morning when wind is high, the excess energy is used in the charging process of CAES, and when electricity price is low the CAES system can also be charged. However, during periods of high demand, both RES and CAES sell energy to the market to maximise their profit. The system operator ensures that a certain percentage of energy dispatched on a daily basis is composed of RESs, such as wind and PV. This is achieved by using the renewable obligation framework which is presented in [8], [9] and [12].

9.3.1   Renewable obligation framework

The main purpose here is to minimise the total operating cost for the system operator thermal generators while maintaining a percentage of renewable energy in the energy mix.

9.3.1.1 Objective functions

The objective is to maximise the profit (τ) of the RES–CAES aggregator participating in the day-ahead market. The profit is defined as the difference between the revenue (τ_1) generated from selling RES–CAES energy and the total operating cost (τ_2).

$$\tau = \tau_1 - \tau_2 \tag{9.1}$$

The total revenue generated is given in (9.2):

$$\tau_1 = \sum_{t=1}^{T} \lambda_t^R \left(\sum_{m=1}^{N_M} p_{m,t}^{\text{Wind}} + \sum_{v=1}^{N_V} p_{v,t}^{\text{PV}} + \sum_{c=1}^{N_C} p_{c,t}^{\text{CAES},d} \right) \tag{9.2}$$

The operating cost of thermal generators and the penalty function are determined using (9.2). Equation (9.2) has three parts: the first part is associated with the thermal generator operating costs, i.e. fixed cost, variable cost, start-up and shut-down cost. These costs are typical for thermal generators and they are associated with unit-commitment model. The second part of (9.2) is associated with CAES charging status; this means whenever the CAES unit charges up and the wind generators cannot provide the energy for charging, then the thermal generators must provide energy for the CAES using market price. The last part of (9.2) is related to the renewable obligation requirement. If the overall energy mix is less than the required renewable obligation, then a penalty is imposed on thermal generators by the system operator [12].

$$\tau_2 = \sum_{t=1}^{T} \sum_{g=1}^{N_G} \left(C_{g,t}^F u_{g,t} + C_{g,t}^V p_{g,t} + C_g^{\text{SU}} y_{g,t} + C_g^{\text{SD}} z_{g,t} \right) + \sum_{t=1}^{T} \sum_{c=1}^{N_C} \lambda_t^R p_{c,t}^{\text{CAES},c} + \Upsilon \tag{9.3}$$

where $C_{g,t}^F$ is the fixed cost of generating unit g in time period t, $C_{g,t}^V$ is the variable cost of generating unit g, $C_{g,t}^{\text{SU}}$ is the start-up cost of generating unit g, $C_{g,t}^{\text{SD}}$ is the shut-down cost of generating unit g at time period t, λ_t^R is the price of electricity and $p_{c,t}^{\text{CAES},c}$ is the charging power of CAES system. Note that $u_{g,t}$ is a binary variable that is equal to 1 if generating unit g is online in time period t and 0 otherwise; $p_{g,t}$ is the output power of generating unit g during time period t; $y_{g,t}$ is the binary variable that is equal to 1 if generating unit g is started up at the beginning of time period t and 0 otherwise; and $z_{g,t}$ is a binary variable that is equal to 1 if generating unit g is shut down at the beginning of time period t, and 0 otherwise.

The notation Υ is the second part of the total cost which is the renewable obligation part of the model shown in (9.4):

$$\Upsilon = \gamma \sum_{t=1}^{T} \left(\alpha \left(\sum_{g=1}^{N_G} p_{g,t} + \sum_{m=1}^{N_M} p_{m,t}^{\text{Wind}} + \sum_{v=1}^{N_V} p_{v,t}^{\text{PV}} + \sum_{c=1}^{N_C} \left(p_{c,t}^{\text{CAES},d} - p_{c,t}^{\text{CAES},c} \right) \right) \\ - \left(\sum_{m=1}^{N_M} p_{m,t}^{\text{Wind}} + \sum_{v=1}^{N_V} p_{v,t}^{\text{PV}} \right) \right)^{+} \tag{9.4}$$

where γ is the penalty imposed to the thermal generators for not achieving the renewable obligation and α is the renewable obligation requirement in percentage. The sign function $(\cdot)^+$ is equal to γ if the RES obligation is unattained and 0 otherwise. The penalty γ is normally provided by the energy regulator as an annual value. This penalty value can be changed to a daily penalty value corresponding to daily economical dispatch of generators.

9.3.1.2   Maximising renewable energy penetration

The second objective function increases the level of RES penetration as presented in [12]. In addition to the total operating cost in (9.3), the maximisation of RES penetration is shown in (9.5):

$$\tau_3 = \sum_{t=1}^{T} \left(\sum_{m=1}^{N_M} p_{m,t}^{\text{Wind}} \Delta t + \sum_{v=1}^{N_V} p_{v,t}^{\text{PV}} \Delta t \right) \tag{9.5}$$

9.3.1.3   Constraints

The constraints for the optimal dispatch of RES and thermal generators without CAES are shown in the following. Constraint (9.6) means that the total power generated from wind, PV and thermal generators is equal to the demand per hour at each load bus. Constraints (9.7) and (9.8) show the maximum allowable ramp rate of the thermal generators. Constraints (9.9) and (9.10) provide the generator logical expression which ensures that any thermal generating unit that is online can be shut down but not started up. In a similar sense, any generating unit that is offline can be started up but not shut down. Constraints (9.11)–(9.13) give the thermal, PV and wind generator capacity limit. While constraint (9.14) is a security constraint that ensures the sum of all thermal generators is always greater than the sum of the demand and requires reserves for the entire dispatch period. Constraints (9.15) and (9.16) represent the transmission flow limits which are estimated using DC power flow.

$$\sum_{g=1}^{N_G} p_{g,t} + \sum_{m=1}^{N_M} p_{m,t}^{\text{Wind}} + \sum_{v=1}^{N_V} p_{v,t}^{\text{PV}} = \sum_{b=1}^{N_B} P_{b,t} \quad \forall t \tag{9.6}$$

$$p_{g,t} - p_{g,t-1} \leq R_g^U u_{g,t-1} + R_g^{\text{SU}} y_{g,t} \quad \forall g, \forall t \tag{9.7}$$

$$p_{g,t-1} - p_{g,t} \leq R_g^D u_{g,t} + R_g^{\text{SD}} z_{g,t} \quad \forall g, \forall t \tag{9.8}$$

$$y_{g,t} - z_{g,t} = u_{g,t} - u_{g,t-1} \quad \forall g, \forall t \tag{9.9}$$

$$y_{g,t} + z_{g,t} \leq 1 \quad \forall g, \forall t \tag{9.10}$$

$$p_{g,t}^{\min} u_{g,t} \leq p_{g,t} \leq p_{g,t}^{\max} u_{g,t} \quad \forall t \tag{9.11}$$

$$p_{m,t}^{\text{Wind}} \leq p_{m,t,\text{gen}} \quad \forall t \tag{9.12}$$

$$p_{v,t}^{\text{PV}} \leq p_{v,t,\text{gen}} \quad \forall t \tag{9.13}$$

$$\sum_{g=1}^{N_G} p_{g,t}^{\max} u_{g,t} \geq P_{b,t} + R_{b,t} \quad \forall t \tag{9.14}$$

$$-P_{l,\max} \leq P_{l,t} \leq P_{l,\max}, \quad \forall l, \forall t \tag{9.15}$$

$$P_{l,t} = \sum_{g=1}^{N_G} G_{l,g} P_{g,t} + \sum_{m=1}^{N_M} F_{l,m} p_{m,t}^{\text{Wind}} + \sum_{v=1}^{N_V} H_{l,v} p_{v,t}^{\text{PV}} = \sum_{c=1}^{N_C} Q_{l,c} \left(p_{c,t}^{\text{CAES},d} - p_{c,t}^{\text{CAES},c} \right)$$

$$- \sum_{b=1}^{N_B} D_{l,b} P_{b,t} \tag{9.16}$$

where $G_{l,g}$, $F_{l,m}$, $H_{l,v}$, $Q_{l,c}$ and $D_{l,b}$ denote the generator shift factor coefficient between line l and thermal generator, wind farms, PV plant, CAES system and system demand at each bus, respectively. The transmission line power $P_{l,t}$ of line l at time interval t is calculated using DC power flow.

9.3.2   CAES mathematical model

The operating cost of the CAES system is already included in (9.3). The technical constraints associated to CAES are expressed in (9.17)–(9.24). Constraint (9.17) shows the amount of energy injected into storage for later usage, while constraint (9.18) gives the amount of energy produced by CAES system. Constraints (9.19) and (9.20) provide the mathematical model of the air stored in storage and then pumped from storage to the expander for energy production. In order to prevent the charging and discharging process from occurring simultaneously, constraint (9.21) is used. Constraint (9.22) shows the dependence of stored capacity from the previous time period to the current dispatch period and constraint (9.23) provides the storage limit of the CAES system. Finally constraint (9.24) ensures that the initial and final storage state is maintained [13].

$$V_{c,t}^{\text{ini}} = \eta_c^{\text{ini}} p_{c,t}^{\text{CAES},c} \quad \forall t, c \tag{9.17}$$

$$p_{c,t}^{\text{CAES},d} = \eta_c^p V_{c,t}^p \quad \forall t, c \tag{9.18}$$

$$V_{\min,c}^{\text{ini}} u_{c,t}^{\text{ini}} \leq V_{c,t}^{\text{ini}} \leq V_{\max,c}^{\text{ini}} u_{c,t}^{\text{ini}} \quad \forall t, c \tag{9.19}$$

$$V_{\min,c}^p u_{c,t}^p \leq V_{c,t}^p \leq V_{\max,c}^p u_{c,t}^p \quad \forall t, c \tag{9.20}$$

$$u_{c,t}^{\text{ini}} + u_{c,t}^p \leq 1 \quad \forall t, c \tag{9.21}$$

$$A_{c,t+1}^{\text{Store}} = A_{c,t}^{\text{Store}} + V_{c,t}^{\text{ini}} - V_{c,t}^p \quad \forall t, c \tag{9.22}$$

$$A_{c,\min}^{\text{Store}} \leq A_{c,t}^{\text{Store}} \leq A_{c,\max}^{\text{Store}} \quad \forall t, c \tag{9.23}$$

$$A_{c,t_0}^{\text{Store}} = A_{c,t_f}^{\text{Store}} \tag{9.24}$$

The introduction of the CAES system affects the power balance constraint (9.6), which must be adapted to include the impact of CAES. The new power balance constraint is shown in (9.25):

$$\sum_{g=1}^{N_G} P_{g,t} + \sum_{m=1}^{N_M} P_{m,t}^{\text{Wind}} + \sum_{v=1}^{N_V} P_{v,t}^{\text{PV}} + \sum_{c=1}^{N_C} \left(P_{c,t}^{\text{CAES},d} - P_{c,t}^{\text{CAES},c} \right) = \sum_{b=1}^{N_B} P_{b,t} \quad \forall t$$

(9.25)

9.3.3 Direct load control model

The price-based demand response objective function presented in (9.26) is known as the customer utility function which aims to minimise the discomfort level due to the lack of electricity.

$$\tau_4 = \sum_{t=1}^{T} \sum_{b=1}^{N_B} \left(\left(\lambda_t^R - \xi_{b,t}^c \right) \left(P_{b,t}(1 - u_{b,t}) + \widetilde{P}_{b,t} u_{b,t} \right) \right) \Delta t$$

$$+ \sum_{t=1}^{T} \sum_{b=1}^{N_B} \Delta P_{b,t} u_{b,t} \xi_{b,t}^i$$

(9.26)

Equation (9.26) refers to the total cost associated with minimising the customer discomfort level which measures the benefit the consumer achieves by using electricity during time period t, λ_t^R is the RTP of electricity and $\xi_{b,t}^c$ is the benefit or willingness of the customer to buy electricity for performing tasks requiring electricity. For simplicity, the benefit for the consumer is assumed to be constant and time independent, which implies that the residential load is deferrable since the task can be performed at any time during the day [14].

To encourage residential customer participation, an incentive is introduced to the model to quantify the impact of demand reduction and demand deferred. As a result of the incentive, the residential customers are incentivised only during peak hours which is assumed to correspond to high demand period.

The incentive price paid to customers is $\xi_{b,t}^i$, and $\Delta P_{b,t}$ is the difference between the actual demand at the participating demand response programme (DRP) bus b before and after the demand reduction.

$$\Delta P_{b,t} = P_{b,t} - \widetilde{P}_{b,t}$$

(9.27)

The direct load control switching status $u_{b,t}$ is a binary variable that is equal to 1 if a DRP, such as a residential load management (RLM) program is implemented at bus b in time t and 0 indicating that no RLM is implemented.

9.3.3.1 Constraints changes

The only changes in the constraints are due to the change in demand which is replaced by (9.28); the constraints affected by the demand reduction are (9.6) and (9.14).

$$P_{b,t} = \sum_{b=1}^{N_B} \left(P_{b,t}\left(1 - u_{b,t}\right) + \widetilde{P}_{b,t}u_{b,t} \right) \quad \forall t \tag{9.28}$$

The two cost functions can be added together to form a new objective function as shown in (9.29):

$$\tau_{TC} = \tau_2 + \tau_4 \tag{9.29}$$

9.3.4   Wind energy system

The intermittent output power of a wind turbine can be characterised as a random variable which is related to the wind speed at the hub of the turbine. The actual intermittent power can be represented as a function of wind speed (9.30) [15]. Moreover, the wind output power can be transformed from wind speed using a statistical transformation given in [16,17].

$$p_{m,t,\text{gen}}\left(\pi_{m,t}\right) = \begin{cases} 0 & \text{if } \pi_{m,t} < \pi_m, \pi_{m,t} > \pi_o \\ p_{m,r}\Gamma(t) & \text{if } \pi_m \le \pi_{m,t} \le \pi_r \\ p_{m,r} & \text{if } \pi_r \le \pi_{m,t} \le \pi_o \end{cases} \tag{9.30}$$

The wind speed $\pi_{m,t}$ is a random variable that varies over time, where π_m, π_r and π_o are the wind turbine cut in speed, rated speed and cut out speed, respectively, all in m/s. This means that the corresponding wind power is also a random variable and $\Gamma(t)$ is shown in (9.31):

$$\Gamma(t) = \left(\frac{\pi_{m,t} - \pi_m}{\pi_r - \pi_m} \right) \tag{9.31}$$

9.3.5   Solar energy system

For a PV energy system, a relationship among radiation resource, temperature and output power can be found in [18], which is also given by the function (9.32):

$$p_{v,t,\text{gen}}\left(\Omega_t\right) = \begin{cases} p_{\text{vr}}(\Omega_t^2/\Omega_{\text{std}}R_c) & \text{if } 0 < \Omega_t < R_c \\ p_{\text{vr}}(\Omega_t/G_{\text{std}}) & \text{if } \Omega_t > R_c \\ 0 & \text{if } G_t = 0 \end{cases} \tag{9.32}$$

where PV cell temperature is neglected, and the solar active power generation can either be controlled by maximum power point tracking algorithm or be charged into batteries. This means that the maximum penetration of the PV generator is limited by the available maximum active power generation which is subject to solar irradiation and temperature [19,20].

9.3.6   Weighted-sum approach

There are generally several methods that can be used to solve a multi-objective optimisation problem in the literature. These methods are typically the weighted sum [21], global criterion [22] and ε-constraint [23] to mention a few. In this chapter, the weighted-sum method is used to change the multi-objective optimisation problem into a single-objective problem. The reason for selecting this method originates from the fact that it is efficient and easy to use for any multi-objective problem. The two objective functions in (9.1) and (9.5) are shown in (9.33):

$$\max J = \vartheta \tau + (1 - \vartheta)\tau_3 \tag{9.33}$$

The weighting coefficient ϑ is selected so that the overall problem is not favouring one objective over the other. This will ensure that the overall optimisation problem is a maximisation. Moreover, the weighting factor is constant over the entire dispatch period. The new objective function (9.33) is subject to constraints (9.6)–(9.25).

9.3.7 Solution approach

The optimisation model presented is a mixed-integer linear programming problem and it is implemented using IBM ILOG CPLEX. The tolerance gap in the CPLEX solver is set to 0.02%. The simulation is conducted on a notebook with an Intel Core i5 at 2.70 GHz and 8 GB RAM. A step-by-step approach for implementing the proposed optimisation model is as follows:

1. Input demand, 24-h RES output data, and day-ahead electricity price.
2. Calculate the forecast RES generation $p_{m,t}^{\max}, p_{v,t}^{\max}$ using (9.30) and (9.32) respectively.
3. Select the starting weighting factor ϑ for the optimisation model (9.33).
4. Solve the optimisation model.
5. Provide the first optimal solution associated with the weighting factor.
6. Provide a bidding strategy for the joint RES–CAES system that ensures maximum profit while meeting a specified renewable obligation.

9.4 Numerical simulation

In this section, the modified IEEE 30-bus system is used to validate the proposed optimisation model over a 24-h time horizon with a temporal resolution of 60 min. In order to demonstrate the effectiveness of the proposed model, four case studies are used in the numerical simulations:

Case 1: Normal participation of RES in the day-ahead market without demand response and CAES.
Case 2: Normal operation with RES and demand response in the day-ahead market.
Case 3: Normal operation with RES and CAES in the day-ahead market.
Case 4: Normal operation with RES, CAES, and demand response in the day-ahead market.

A comparison of the four case studies is then used to show the difference and benefits of using a CAES system in the operation of RES generators.

9.4.1   Input data

The modified IEEE 30-bus system has 6 thermal units and 41 transmission lines. The system data for IEEE 30-bus can be found in [24]. In all the simulation studies, an RES penetration level of 10% is used as a benchmark and the weighting factor is set to 0.5. In the case where RES penetration level is not achieved, a penalty of $100,000 per day is imposed [9]. The parameters for thermal generators are given in Table 9.1 and the parameters of the CAES system are taken from [13] and presented in Table 9.2.

The modified IEEE 30-bus test system has a total demand of 26,822.50 MW before direct load control and a total demand of 26,501.30 MW after the implementation of direct load control of electric water heaters at a substation level. The integrated PV and wind farms are connected to buses 7, 15, 22 and 24, i.e. two PV plants and two wind farms. The sizes of the PV and wind farms are 75, 140, 300 and 500 MW with an installed capacity of 1,015 MW. The CAES system is connected to buses 22 and 24 together with the wind farms. The direct load control is implemented at a substation level, and the substations participating in the

Table 9.1   Thermal generator parameters

Parameter	Unit 1	Unit 2	Unit 3	Unit 4	Unit 5	Unit 6
RD (MW/h)	60	60	60	60	60	60
RU (MW/h)	60	60	60	60	60	60
P_{min} (MW)	50	50	50	50	50	50
P_{max} (MW)	350	250	150	350	450	500
SD (MW/h)	60	60	60	60	60	60
SU (MW/h)	60	60	60	60	60	60
FC ($)	5	7	6	4	5	8
SU ($)	9	12	12	15	14	18
SD ($)	0.5	0.9	1	0.2	1.2	0.8
VC ($)	7	10	8.	11	10.5	12
RUC ($/MWh)	1.6	1.3	1.2	1.1	1.0	0.9
RDC ($/MWh)	1.5	1.2	1.1	1.0	0.9	0.8

Table 9.2   CAES system parameters

Parameter	Unit 1	Unit 2
$V_{min,c}^{ini}$ (MWh)	10.0	10.0
$V_{max,c}^{ini}$ (MWh)	50.0	50.0
$V_{min,c}^{p}$ (MWh)	10.0	10.0
$V_{max,c}^{p}$ (MWh)	50.0	50.0
$A_{c,min}^{store}$ (MWh)	40.0	40.0
$A_{c,max}^{store}$ (MWh)	200	200
η_c^{ini} (%)	0.80	0.95
η_c^{p} (%)	0.80	0.95

Table 9.3   *PV solar irradiance profile for Sites 1 and 2*

Parameter	PV 1	PV 2
R_c (W/m^2)	150	150
Ω_{std} (W/m^2)	900	980

Table 9.4   *Wind speed profile for Sites 1 and 2*

Parameter	Wind 1	Wind 2
π_m (m/s)	2.5	3.0
π_r (m/s)	10	12
π_o (m/s)	18	20

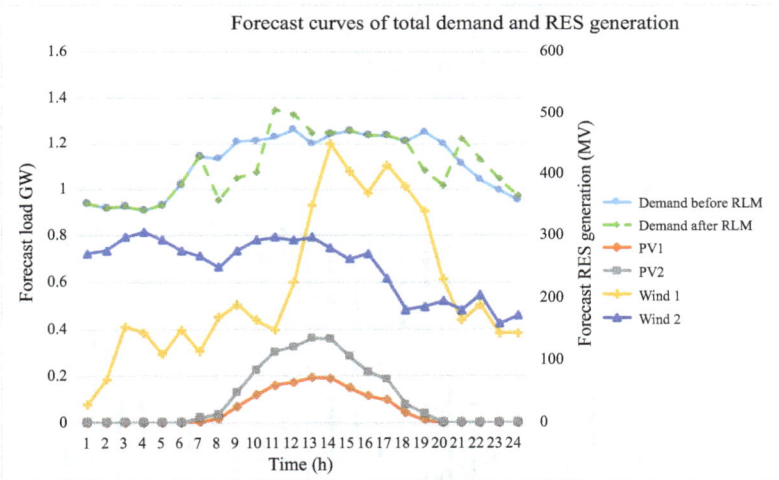

Figure 9.2   Forecasted demand and RES generation before and after direct load control of electric water heaters at a substation level

programme are located on buses 10, 14, 23 and 26. The sizes of the CAES system are 10 and 30 MW [13]. The parameters of RES generators are listed in Tables 9.3 and 9.4, respectively [12].

The total demand before and after the implementation as well as the forecasted RES penetration is shown in Figure 9.2 and the electricity price is shown in Figure 9.3.

9.4.1.1   Case 1

In the normal operating condition, the thermal generators contribute 67.63% and RES generators contribute 32.37% which is more than the minimum renewable

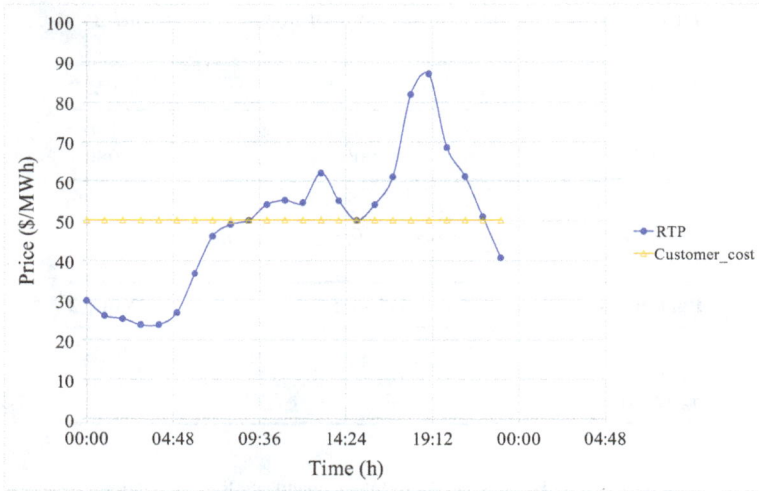

Figure 9.3   RTP of electricity and the customer price of electricity

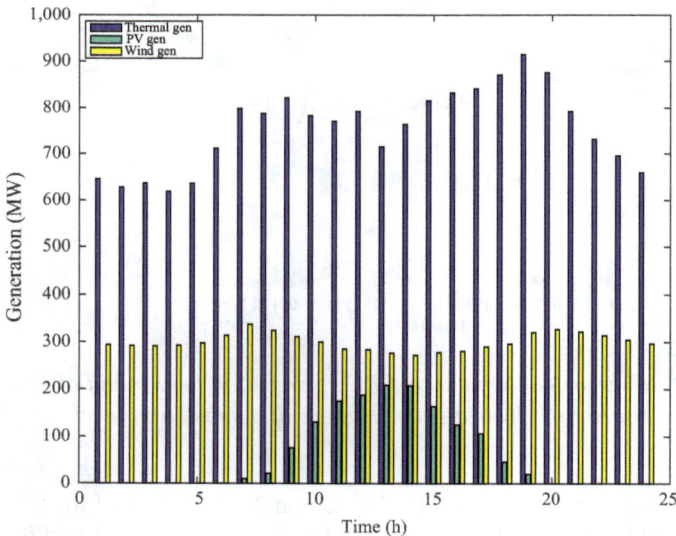

Figure 9.4   Normal day-ahead dispatch schedule for RES and thermal generators

obligation requirement of 10%. This means that no penalty is imposed for the combined operation of RES and thermal generators. The total operating cost is calculated from (9.3); however, the CAES operating cost is excluded. Therefore, a total profit of $281,591.21 is achieved. Figure 9.4 shows the total generation of thermal and RES generators.

The generation from thermal units makes up the largest contribution, which is followed by RES generators. The total operation costs for supplying electrical demand is $314,520. The thermal units generate 18,581.50 h (69.28%) and RES generators contribute 8,241.01 MWh (30.72%) to the total demand. A maximum of 8,456.80 MW is allocated for spinning reserve services.

9.4.1.2   Case 2

In this case study, demand response is added to the existing model to determine the profit as well as the RES penetration. The total operating cost is $153,030, which is calculated from (9.3) and excludes the CAES cost term. This cost includes the demand response incentive cost which is calculated from (9.26), while the total revenue is $434,150 and the profit achieved is $295,605.15. The achieved demand reduction is 375.47 MW, which is 1.4% of the total demand. The RES penetration is 32.81% which is slightly higher than the base case. The base load is supplied by thermal generators which contribute 67.193% and is lower than the base case. The impact of including demand response to the proposed model increases the profits. Figure 9.5 shows the day-ahead dispatch schedule for thermal and RES generators with demand response.

9.4.1.3   Case 3

In this case study, the CAES storage is introduced to the model and the total operating cost is calculated from (9.3); however, demand response is not implemented. The achieved profit for this case study is $281,387.16, which is slightly lower than the base case and the reason for such a lower profitability is due to the

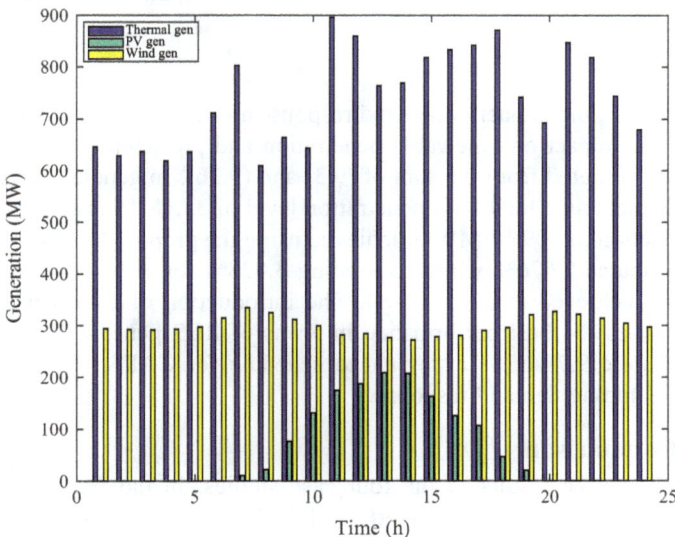

Figure 9.5   Day-ahead dispatch schedule for RES and thermal generators under demand response

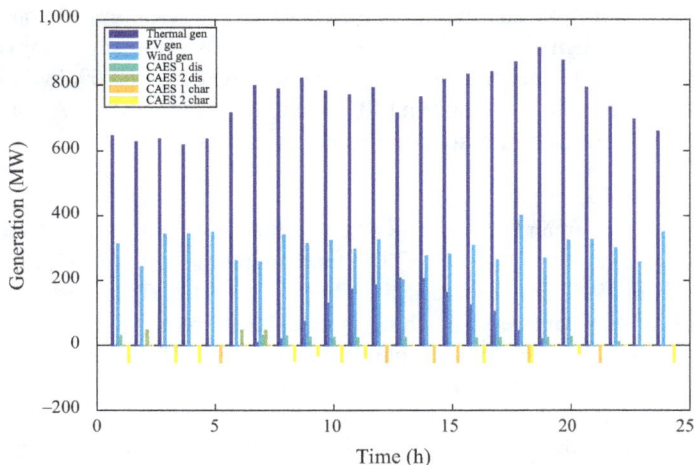

Figure 9.6   Day-ahead 24-h schedule with CAES, RES and thermal generators

charging process of the CAES system which uses electricity from the market for charging purposes and as a result increases the operating cost. The total RES penetration achieved is 32.713%, which is higher than the base case. The CAES system contributes 2.723% to the total demand and uses 3.079% for charging purposes. The base load is supplied by thermal generators which contribute 67.643% of the total energy. Figure 9.6 shows the dispatch schedule for a 24-h period.

9.4.1.4   Case 4

In this case study, the impact of demand response and CAES system is investigated to measure the impact on renewable penetration and profitability. The total operating cost is computed from the sum of (9.3) and (9.26). In general, the renewable obligation is met and a total RES penetration level of 33.183% is achieved. A total demand reduction of 309.67 MW is achieved using the proposed model. The CAES system contributes 2.676% while consuming 3.028% for charging purposes. The overall profit achieved is $295,635.84. The introduction of CAES with demand response realises the best performance in terms of RES penetration and demand reduction. Figure 9.7 shows a day-ahead schedule for CAES, RES and thermal generators including demand response.

9.4.1.5   Comparison

Table 9.5 shows the results of the four case studies. In the first case, the total demand is supplied by the combination of RES and thermal generators, and a maximum profit of $281,591.21 is achieved. There is no demand response participation. The total RES penetration achieved is 32.37%. In Case 2, the profit is increased compared to the first case study; there is also a slight increase in RES

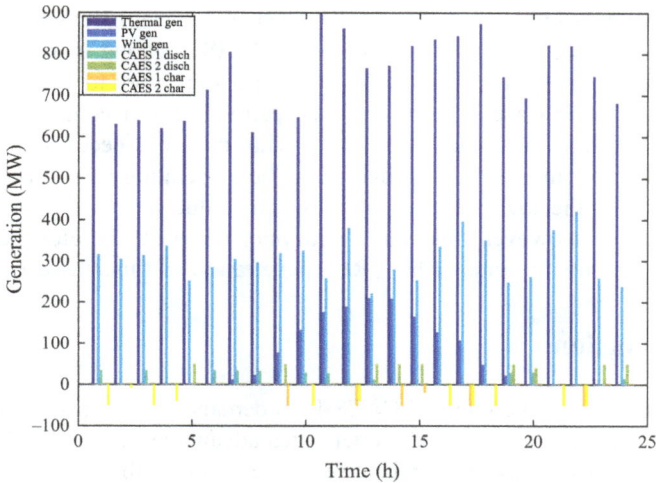

Figure 9.7   Day-ahead schedule with demand response and CAES, RES and thermal generators

Table 9.5   Case study comparison

Description	Case 1	Case 2	Case 3	Case 4
Profit ($)	281,591.21	295,604.15	281,387.16	295,625.84
Total cost ($)	152,848.79	138,545.85	192,792.84	178,984.16
Revenue ($)	434,440	434,150	474,180	474,610.00
DR (MW)	0.00	375.47	0.00	309.67
Total demand (MW)	26,822.46	26,447.00	26,822.46	26,420
RES pen (%)	32.37	32.81	32.71	33.18
Gen pen (%)	67.63	67.19	67.64	67.17
CAES discharge (%)	0.00	0.00	2.72	2.68
CAES charging (%)	0.00	0.00	3.08	3.03

penetration. In addition, there is a demand reduction of 375.47 MW; this means that the proposed model takes advantage of the flexible load by deferring it to periods of low energy cost, and as a result, this reduces the total operating costs compared to the first case.

A comparison of Cases 3 and 4 shows that there is a significant increase in RES penetration when CAES and demand response is combined in the day-ahead dispatch schedule. The impact of adding demand response helps in reducing the overall operating costs while the impact of adding CAES system assists in increasing the RES penetration which is shown by an increase in RES penetration.

A comparison of the different case studies shown in Table 9.5 gives different profits for different cases. Although the profit margin between Cases 2 and 4 is small, it should be noted that including the CAES and demand response (DR) will have more benefits in terms of the overall revenue. For example, when CAES and DR are implemented as shown in Case 4, the CAES charging increases the operating cost significantly since the thermal generators are used to charge the CAES system. However, in Case 2, the revenue is relatively lower compared to Case 4, which is due to the lower operating costs. This does not necessarily mean that demand response without CAES provides the best operating mode. The only benefit in Case 2 is the low operating cost, which is due to the implementation of demand response; however, this comes with poor revenue. Therefore, Case 4 provides better benefits in terms of high RES penetration and increased revenue.

9.5   Conclusion

This chapter proposes a combined CAES-RES demand response programme using direct load control. The proposed model takes advantage of the CAES system to deal with the uncertainty of wind power generation in the day-ahead dispatch schedule. The proposed model is implemented on a standard IEEE 30-bus system and the effectiveness of the model is tested on four case studies. The inclusion of CAES system increases RES penetration and improves total profit using the IEEE 30-bus system. The implementation of storage and direct load control at the substation level affects the total operating costs as illustrated in the comparative study.

1. Using the modified IEEE 30-bus, the proposed optimisation model can increase RES penetration and participate in the day-ahead energy market with minimal uncertainty.
2. The implementation of CAES energy system improves profitability and reduces the total operating cost.
3. The CAES system can improve the operability of RES generators and assist them in the day-ahead market participation.

The proposed model does not take into consideration the uncertainty associated with wind and PV system. Therefore, the future research will incorporate that into the model while including the secondary market for renewable obligation certificates.

Acknowledgements

This work was supported in part by the South African National Energy Development Institute.

References

[1]   He W, Dooner M, King M, *et al.* Techno-economic analysis of bulk-scale compressed air energy storage in power system decarbonisation. *Applied Energy.* 2021;282:116097.

[2] Cleary B, Duffy A, O'Connor A, *et al.* Assessing the economic benefits of compressed air energy storage for mitigating wind curtailment. *IEEE Transactions on Sustainable Energy.* 2015;6(3):1021–1028.

[3] Attarha A, Amjady N, Dehghan S, *et al.* Adaptive robust self-scheduling for a wind producer with compressed air energy storage. *IEEE Transactions on Sustainable Energy.* 2018;9(4):1659–1671.

[4] Shafiee S, Zareipour H, Knight AM, *et al.* Risk-constrained bidding and offering strategy for a merchant compressed air energy storage plant. *IEEE Transactions on Power Systems.* 2016; 1–1.

[5] Khatami R, Oikonomou K and Parvania M. Look-ahead optimal participation of compressed air energy storage in day-ahead and real-time markets. *IEEE Transactions on Sustainable Energy.* 2020;11(2):682–692.

[6] Ghadi MJ, Azizivahed A, Rajabi A, *et al.* Day-ahead market participation of an active distribution network equipped with small-scale CAES systems. *IEEE Transactions on Smart Grid.* 2020;11(4):2966–2979.

[7] Ghavidel S, Ghadi MJ, Azizivahed A, *et al.* Risk-constrained bidding strategy for a joint operation of wind power and CAES aggregators. *IEEE Transactions on Sustainable Energy.* 2020;11(1):457–466.

[8] Hlalele TG, Naidoo RM, Bansal RC, *et al.* Multi-objective stochastic economic dispatch with maximal renewable penetration under renewable obligation. *Applied Energy.* 2020;270:115120.

[9] Hlalele TG, Zhang J, Naidoo RM, *et al.* Multi-objective economic dispatch with residential demand response programme under renewable obligation. *Energy.* 2021;218:119473.

[10] Vieira FS, Balestieri JAP and Matelli JA. Applications of compressed air energy storage in cogeneration systems. *Energy.* 2021;214:118904.

[11] Tong Z, Cheng Z and Tong S. A review on the development of compressed air energy storage in China: Technical and economic challenges to commercialization. *Renewable and Sustainable Energy Reviews.* 2021;135:110178.

[12] Hlalele TG, Naidoo RM, Zhang J, *et al.* Dynamic economic dispatch with maximal renewable penetration under renewable obligation. *IEEE Access.* 2020;8:38794–38808.

[13] Ghalelou AN, Fakhri AP, Nojavan S, *et al.* A stochastic self-scheduling program for compressed air energy storage (CAES) of renewable energy sources (RESs) based on a demand response mechanism. *Energy Conversion and Management.* 2016;120:388–396.

[14] Marales JM, Conejo AJ, Madsen H, *et al.*, editors. *Integrating Renewables in Electricity Markets.* Berlin: Springer; 2014.

[15] Bansal RC, editor. *Handbook of Distributed Generation: Electric Power Technologies, Economics and Environmental Impacts.* Berlin: Springer; 2017.

[16] Patel MR. *Wind and Solar Power Systems: Design, Analysis and Operation.* 2nd ed. Boca Raton, FL: Taylor & Francis; 2006.

[17]   Manwell JF, McGowan JG and Rogers AL. *Wind Energy Explained: Theory, Design and Application*. 2nd ed. New York, NY: Wiley; 2009.

[18]   Liang R and Liao J. A fuzzy optimization approach for generation scheduling with wind and solar energy systems. *IEEE Trans Power Syst*. 2007; 22(4):1665–1674.

[19]   Lupangu C and Bansal RC. A review of technical issues on the development of photovoltaic systems. *Renewable & Sustainable Energy Reviews*. 2017;73:950–965.

[20]   Nghitevelekwa K and Bansal RC. A review of generation dispatch with large-scale Photovoltaic (PV) systems. *Renewable & Sustainable Energy Reviews*. 2018;81:615–624.

[21]   Harkouss F, Fardoun F and Biwole PH. Multi-objective optimization methodology for net zero energy buildings. *Journal of Building Engineering*. 2018;16:57–71.

[22]   Kianmehr E, Nikkhah S and Rabiee A. Multi-objective stochastic model for joint optimal allocation of DG units and network reconfiguration from DG owner's and DisCo's perspectives. *Renewable Energy*. 2019;132:471–485.

[23]   Nasr MA, Nikkhah S, Gharehpetian GB, *et al.* A multi-objective voltage stability constrained energy management system for isolated microgrids. *International Journal of Electrical Power & Energy Systems*. 2020;117:105646.

[24]   Hu F, Hughes KJ, Ingham DB, *et al.* Dynamic economic and emission dispatch model considering wind power under energy market reform: A case study. *International Journal on Electrical Power & Energy Systems*. 2019;110:184–196.

Chapter 10

Direct air capture and wind curtailment: a technology-based business approach for the US market

Erik R. Steeb[1] and George Xydis[1,2]

As the grids are continually modernized and integrate higher amounts of renewable energy sources, further deployment of renewables could bring further abrupt reduction in wind-based electricity generation, also known as wind curtailment. The electricity markets are able to economically and technically lay the foundations to facilitate further penetration (by massively offering negative prices via demand response); however, since independent power producers (and subsequently wind manufacturers) need a stable growth and profitable plan not dependent on each market's marginal prices, the pressure from the industry on establishing another system based on negative emissions will also grow. Direct air capture (DAC) could play that role in the immediate future. It is widely accepted that curtailment will continue to be present, but in order its rate to fall, investments are needed, especially in the markets where wind farms are not meeting the majority of electricity demand, and there is still a long way, such as the Chinese and the US markets. In this work, a business and technology-based analysis for the US market points out the political decisions required for such a long and decisive step.

Keywords: Direct air capture; wind curtailment; technology-based business solution

10.1 Introduction

The consistent increase of wind and solar electricity in the US has been a major driver of the decrease in CO_2 emissions from the electricity sector over the past 15 years. From 2005 to 2018, emissions associated with electricity production

[1]Energy Policy and Climate Program, Krieger School of Arts and Sciences, Johns Hopkins University, Baltimore, MD, USA
[2]Department of Business Development and Technology, Aarhus University, Birk Centerpark, Herning, Denmark

declined nearly 30%, with about a quarter of that reduction from replacing fossil fuels with non-carbon sources of electricity, predominately wind and solar [1]. This has come in part due to increased environmental considerations at both state and federal levels, but has predominately been driven by consistent cost declines (increasing, at the same time, power output), especially for wind [2]. By late 2019, unsubsidized onshore wind had the least expensive levelized cost of energy (LCOE) for new generation – and even beat out some instances of continued operation of existing nuclear and coal plants [3,4].

The rapid increase in renewables on the electric grid has also presented some challenges for the management of the whole electric grid, especially in terms of the relationship between various energy sources, such as intermittent sources, battery and hydrogen storage [5,6]. During peak periods of both solar and wind production, demand may not be high enough for the total production, and renewable energy can be curtailed off the grid [7]. While curtailment of wind is a less prevalent problem than the curtailment of solar, especially on the American grid, it still occurs and reduces revenues for wind farm operators. Wind development is still profitable at current rates of curtailment, and likely will be even at increased levels, but any curtailment still presents a loss of potential revenue for wind operators [8].

While rapid deployment of renewable energy is making an important contribution to reducing current emissions levels, these measures are unlikely to be sufficient to limit global temperature rise to 1.5 °C. The Intergovernmental Panel on Climate Change's (IPCC's) 2018 report 'Mitigation Pathways Compatible with 1.5 °C in the Context of Sustainable Development' found that 'all analysed pathways limiting warming to 1.5 °C with no or limited overshoot use [Carbon Dioxide Removal] to some extent' [9].

One promising type of carbon dioxide removal (CDR) is DAC, where atmospheric CO_2 is removed from the air and sequestered, generally into deep geologic formations, including depleted oil wells. CO_2 captured from the atmosphere can also be used for enhanced oil recovery, where it is injected into active wells in order to squeeze more oil from the well, but potential revenues from additional oil recovery will not be considered here because of the broad environmental harms of the continued oil production.

Because of how to diffuse atmospheric CO_2 is – 416 ppm [10] is only 0.0416% of the atmosphere – any DAC technology is highly energy intensive. Most traditional DAC implementations run air through a solvent that absorbs CO_2 from the atmosphere and then must be heated to release pure CO_2, which is then captured and sequestered. Powering DAC with electricity directly from the grid may be prohibitively expensive, especially given the relatively small potential income streams for CDR broadly and for DAC (which generally has no marketable by-products) specifically. As a result, DAC installations need to minimize their variable input costs. Some proposals currently exist to use waste heat from industrial processes to provide the low-grade heat required to regenerate some novel types of CO_2 solvent, but do not address electricity input requirements [11].

By using electricity from wind projects that would otherwise be curtailed due to an oversupply of renewable energy on the electric grid, DAC installations collocated with wind farms could make use of zero-cost electricity in order to remove CO_2 from the atmosphere while providing wind operators a source of income for electricity that would not be sold on the grid. That income can either be in the form of the 45Q tax credit for carbon capture and sequestration or from companies, such as Microsoft, which have committed to funding carbon negative initiatives in order to further reduce their environmental impact [12].

10.2 Literature review

The phenomenon of curtailment is fairly well studied and is never far from the broader discussion of the place of variable renewables, storage, and demand response on the electric grid. It is of enough public interest that the California Independent System Operator (ISO) reports 'wind and solar curtailment by day' and presents eight solutions for managing the oversupply of variable renewable energy on its grid [13].

The National Renewable Energy Laboratory (NREL) produced a comprehensive report by Bird *et al.* [14] on the extent of current wind curtailment throughout the US. Most notably, it highlights the very high level of wind curtailment in Texas from 2008 to 2011, peaking at 18% in 2009. Those issues were able to be resolved primarily by 'transmission expansion and a market redesign to LMP pricing and faster schedules that improved overall system operations' [14]. The report generally highlights increased transmission capacity, which allows wider geographic distribution of wind electricity, and more granular markets, which allow fossil resources to ramp up or ramp down as needed, as the primary drivers of decreasing or managing wind curtailment across most US electricity markets.

Wind curtailment is not a problem unique to the US grid. In fact, the Chinese grid suffers from higher rates of curtailment than anything imaginable in the US [15,16]: 'In 2016, the wind electricity curtailment in China amounted to 49.7 TWh, enough to cover the total annual electricity consumption of Bangladesh, with a population of 163 million' [17]. In the US as in China, some amount of this curtailment problem is due to the highly fragmented nature of the grid itself; local surpluses of electricity cannot be easily transmitted to geographically distant areas that may be able to use the electricity. In China's particular case, lack of coordination results in transmission constraints, system imbalances, as well as overcapacity, ultimately leading to wind curtailment, with no clear interprovincial system for alleviating any of these issues [17,18].

As variable renewables become increasingly more prevalent on the US grid, curtailment is also likely to increase. Jorgenson *et al.* [19] find that, without significant improvements to national transmission capacity, up to 15% of all renewable electricity produced each year could be curtailed. Many of the required transmission improvements to reduce that curtailment by half have already been proposed, but not yet built, often because they cross state lines and there is no clear process for permitting long distance transmission on the scale required here [19]. Jorgenson *et al.* found that, even with about triple the amount of proposed long-distance transmission, just below 5% of renewable electricity will be curtailed. While this is only a third of the curtailment of the reference case, it would still represent hundreds of gigawatt-hours each year for which renewable energy developers would not be paid, although, for wind projects specifically, wear and tear would still accumulate on machines. However, it goes beyond that. Since modern wind energy business is changing and the profitability of investors is clearly more linked to long-term service contracts than the number of wind turbines sold, it ultimately has an impact on the manufacturers' profitability and growth planning [20]. The business viability is also unavoidably linked to the electricity markets and electricity pricing and one could say with price forecasting. If curtailment rate is high, new plants should be activated in the market, which influences the duration of their operation and thus the market participants. This is directly correlated with the price incentives offered and price forecasting [21]. It could even have eventually an impact on natural gas markets and heating markets [22,23].

As discussed above, some amount of CDR will be required by 2050 and remain in continuous operation in order to offset emissions that cannot be avoided and to draw down any excess CO_2 in the atmosphere if emissions targets are overshot [9, p. 96]. Furthermore, the longer countries delay significant action to rapidly reduce greenhouse gas emissions, beyond even the goals of the Paris Climate Accords, the larger the required amount of CDR will be [9]. The more CDR is required, the broader the range of CDR technologies will be needed to build out sufficient capacity. This means that progressively the more and more complex iterations of CDR will need to be deployed, the longer emissions will continue at high rates.

CDR can come in a variety of implementations. The most well-understood are afforestation and reforestation – planting trees, which removes CO_2 steadily as they grow. The new or newly restored forests are then long-term carbon sinks for as long as they are preserved, requiring continued monitoring and conservation efforts [9,24].

Two other technologies, biochar production and bioenergy with carbon capture and sequestration (BECCS), take the biological removal of CO_2 of afforestation/reforestation and combine it with further processing. In case of biochar, biomass is pyrolysed, slowly burned in a way that retains the carbon, until it is 50%–80% carbon by mass [25]. The biochar is then mixed into soil as an additive and can either be used on farmlands to enhance crop yields or to aid in afforestation/reforestation projects [26]. BECCS utilizes a mixture of traditional biomass energy technology and carbon capture and sequestration technology to capture the emissions from biomass, sequestering the carbon in underground geologic formations [27]. Both BECCS and biochar as a soil amendment are included in IPCC pathways as viable technologies. Low and Schafer [27], however, raise significant concerns about how easily these technologies can be scaled, particularly due to the fact that biomass grown for sequestration purposes will inherently compete for land, water, and fertilizer with food crops needed to sustain a growing population.

DAC, by contrast, skips biological removal entirely and uses purely artificial means to power the entire process. This means that DAC has the highest required direct energy inputs of the CDR implementations, but also means that DAC is not subject to the same land, water, and fertilizer constraints of any of the biologically based implementations. DAC is, however, much more capital and resource intensive than other CDR technologies. One of the clearest limitations on deployment is the 'high-grade (900 °C) heat demand of aqueous solution-based DAC' required once the CO_2 has been captured by a solvent in a DAC plant, that solvent must be regenerated to release the CO_2 for permanent storage and so that the solvent can be reused to capture more CO_2 [28]. Some newly developed solvents, using amine chemicals instead of hydroxides, require somewhat less total thermal energy but, more importantly, require much lower absolute temperatures (in the 85–120 °C range), making them more feasible for widespread deployment without relying on burning natural gas to generate heat (Figure 10.1) [11].

DAC is currently in its infancy as a technology, meaning that there is likely significant potential for cost reductions in a learning curve as firms develop more efficient ways of producing and deploying DAC projects. Fasihi *et al.* [28] predict that this type of learning curve could reduce total DAC costs by 50% by 2030, with costs falling to around 15% of current levels by 2050. Realmonte *et al.* [11] came to similar conclusions, especially when considering potential efficiency gains in building amine-type DAC. Ranjan and Herzog [29], however, argue that the cost of the technology itself will not be the primary barrier to deployment, but rather the energy costs of separating highly diffuse CO_2 from the atmosphere. Their concerns about these energy costs lead to a recommendation to proceed fully with biologically based CDR technologies that concentrate carbon or CO_2 before attempting to capture it, like biochar or BECCS.

Because funding for emissions reduction, clean energy, and CDR is not only limited, but often linked, different technologies are almost always in direct

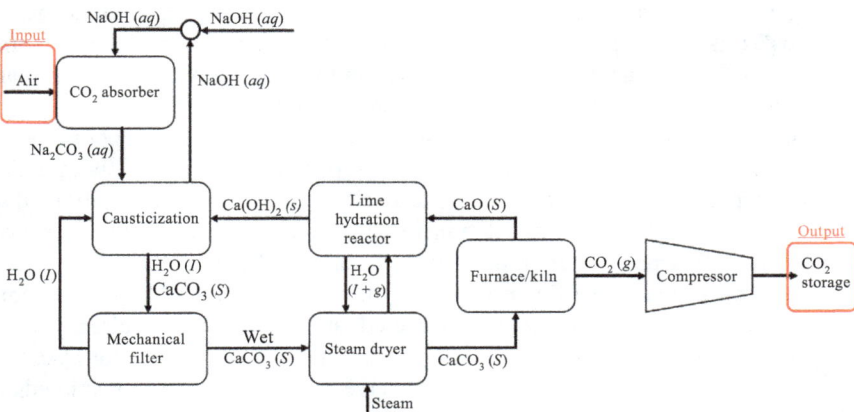

Figure 10.1   DAC process flow diagram using caustic soda

competition for funding. As a result, from an environmentally optimal perspective, funding should go towards whatever marginal additional technology that decreases or avoids the highest amount of emissions. Jacobson explicitly analyses this trade-off, finding that there are a wide variety of potential issues with DAC and that renewable electricity installation (especially wind) has a large, consistent marginal benefit. 'In fact', he concludes 'there is no case where wind powering a synthetic direct air carbon capture and utilization (SDACCU) plant has a social cost below that of wind replacing any fossil fuel or bioenergy power plant directly' [30].

Until the electric grid consists entirely of zero or negative emissions generation, any dollar that could be put towards either decreasing grid emissions or capturing CO_2 from the air is best spent decarbonizing the grid. This generally makes raising capital for DAC potentially more difficult, even if other revenue numbers work well for a specific project.

Finally, there are relatively limited options for generating revenue from DAC. The US currently offers a tax credit that is gradually rising to $50 per metric ton of CO_2 sequestered in 2026, which is only available to facilities sequestering at least 100,000 metric tons per year [31]. While there are some industrial processes that require CO_2, they generally involve the ultimate release of the gas during the process, meaning that additional emissions are avoided, but CO_2 is not completely removed from the atmosphere over an appreciable timescale. This makes corporate funding for DAC relatively hard to come by, although Microsoft has recently pledged to spend money directly on negative emissions [12].

10.3   A meaningful proposal

So far, in many countries where renewables meet a large amount of the energy requirements, significant curtailment rates can be avoided by using negative prices and introducing such incentives [32]. However, since renewables will ultimately continue to occupy more electrical space in the grid – phasing out fossil fuel-based power generation – the more a sustainable solution should be found for 'negative emissions'. It is absolutely understandable now that the market can cope with some negative prices throughout the year; however, on the long term a significant share of negative prices will not be withstood. In order to minimize the total amount of electricity a given wind project has to waste due to curtailment, this paper proposes the collocation of DAC equipment with wind projects that are frequently curtailed in order to provide a demand response capacity at the site of generation, thus eliminating the need for additional transmission capacity beyond what the grid requires in the first place. Action on-site has been also the suggested solution for, e.g., hydrogen production [33] in order for the curtailment problem to be confronted. Energy storage on-site has attracted attention and has already been implemented in smaller (pilot) and larger projects lately. This concept for DAC on-site will work even in locations that do not have the right geology for immediate subterranean storage of CO_2 because the purified compressed gas is fairly easy to inexpensively store, transport, and sequester elsewhere [34].

This type of local demand response is likely necessary even with buildouts of transmission and storage. Although additional transmission investment can reduce the need for curtailment, even 'copper sheet' scenario described by Bird *et al.* [14], where there is infinite transmission capacity between any two points, cannot fully eliminate the need for curtailment – and thus, potential lost revenue. Additionally, storage, both collocated with wind farms and on the grid level, may further reduce curtailment, but it still suffers from limited capacity and potentially from time restrictions (for example, excess wind production could occur over days or weeks, but storage may only operate on a day-to-day charge/discharge cycle).

This proposal operates on the basis that, regardless of increased deployment of transmission and storage, there will be some level of curtailment of wind electricity on an electric grid dominated by variable renewable electricity. Additionally, it operates on the basis that there is no current use for currently curtailed electricity, and that that lack of use reduces potential profits of any wind project that has its production regularly curtailed from the electric grid.

By financing DAC through avoided curtailment of electricity, this model also avoids resource competition between building new renewable energy or CDR. Instead, it opens up a new stream of financing for CDR entirely while potentially making some marginal wind electricity installations profitable where they otherwise would not have been. Additionally, the high capital costs of DAC are unlikely to make it sufficiently profitable for wind developers to be incentivized to build projects strictly for DAC purposes. Transmission and storage investments are also likely more profitable than DAC, preserving those as preferable options before DAC is considered to abate curtailment.

10.4   Further analysis

Table 10.1 synthesizes the energy input requirements in Realmonte *et al.* [11] with the LCOE reported by Lazard [3] in its comprehensive 2019 report on the costs of

Table 10.1   Energy input requirements on the costs of new electricity generation [3,11]

Technology	High heat/electricity			Low heat/electricity		
	Curtailed	Wind, low	Wind, mid	Curtailed	Wind, low	Wind, mid
Hydroxide solvent – high	$300.00	$376.82	$412.49	$300.00	$351.22	$375.00
Hydroxide solvent – low	$180.00	$256.82	$292.49	$180.00	$231.22	$292.49
Hydroxide solvent – min	$100.00	$176.82	$212.49	$100.00	$151.22	$175.00
Amine solvent – high	$350.00	$358.56	$444.30	$350.00	$388.79	$406.80
Amine solvent – low	$200.00	$264.40	$294.30	$200.00	$238.79	$256.80
Amine solvent – min	$50.00	$114.40	$144.30	$50.00	$88.79	$106.80

new electricity generation. In order CDR to be achieved, when air is in contact with some kind of solvent or other chemical media (e.g. hydroxide or amine solvent) is stripped of CO_2 via (usually) heat, ending up to a CO_2 stream to be compressed compression. Even in the lowest heat and electricity input requirement combination (amine solvents), using curtailed wind electricity (with an assumed cost of $0 per MWh) yields a savings of nearly $39 per metric ton of CO_2 sequestered. The high, low, and minimum lines refer to high, low, and minimum costs for capital and labour inputs, per metric ton of CO_2 sequestered.

While capital and labour costs still represent the bulk of the costs in most listed iterations, using zero cost electricity that would otherwise have been curtailed provides significant cost savings, especially in the higher heat and electricity input scenarios. If DAC is able to use zero cost electricity, the marginal inputs required to sequester each ton also have no effect on the total cost of sequestering one ton of CO_2. This can then help drive research and development specifically towards reducing capital and labour costs, concentrating on the technology-related advancements.

Current potential revenue from the 45Q tax credit ($50 per metric ton of CO_2 sequestered) only allow for breaking even in the most optimistic case for amine solvents, showing that, without additional funding, the current iteration of the 45Q tax credit is too low to make DAC profitable on its own.

Revenue from corporations to support negative emissions, however, can bolster the low revenue available from the 45Q tax credit and make the system work. Microsoft does not list a specific number it is willing to pay per ton of CO_2 sequestered, but Stripe, which has made a similar carbon negative pledge, has indicated that it anticipates to initially pay in excess of $100 per ton of CO_2 sequestered [35]. If DAC capital and labour costs can be reduced to the anticipated $50 minimum, additional revenue of $100 per ton would result in additional revenues between $30 and $70 per megawatt-hour devoted to powering DAC – more than may be available from selling the electricity to the grid in the first place. This may be a high estimate for the availability of funding from corporations, but as more firms look towards negative carbon goals, this could be an important source of potential funding.

10.5   Discussion

DAC is heavily energy intensive and the costs associated with those energy requirements can result in the technology is prohibitively expensive. By using electricity produced by wind turbines during times when those turbines would otherwise be curtailed in order to avoid overloading grid resources, operators of wind projects can realize some revenue that would otherwise not exist. The existing statutorily available revenue from the 45Q tax credit is a good start, but wind operators going this route should market their sequestration capacities to companies like Microsoft and Stripe that have made commitments to be carbon negative.

Integrating DAC directly on-site at wind projects is certainly not a simple task – not only would the electrical infrastructure need to be built, but either a pipeline infrastructure for concentrated CO_2 needs to be developed or captured CO_2 needs to be stored on-site until it can be removed periodically by trucks. Either of those options would be logistically difficult undertakings, but, if deployed at large enough scale, not prohibitively expensive or sufficiently difficult as to derail the concept.

Developers of wind projects should consider the addition of DAC at both planned and existing wind developments that either is currently facing significant curtailment or that may be curtailed to a significant degree in the future. This can help those developers ensure that they receive some revenue for each megawatt-hour of electricity generated by their installations, regardless of whether the grid is able to support the additional generation.

In order to ensure further buildout of DAC, Congress should also raise the 45Q tax credit immediately to $100 per metric ton of CO_2 sequestered. The credit should remain at a steady level for at least 10 years before phasing out by 10% of the credit amount per year. Because DAC will likely always require some level of public support, the phase-out should not be total, but the final level needed to support continuous operation of DAC installations should be determined once the long-term capital costs and variable costs are clear. By setting the credit at a constant for a decade, though, and with a predictable start to the phase-out schedule, firms will be able to perform rigorous analyses of potential projects in a stable regulatory environment, instead of having to guess about future revenues. Additionally, Congress should immediately remove the existing requirement that, in order to qualify for the 45Q tax credit, facilities need to sequester at least 100,000 tons of CO_2 per year. This requirement effectively removes the ability of many smaller installations to receive revenue from sequestration, and their aggregate effects on atmospheric emissions could be significant.

10.6   Conclusion

No matter how smart our networks will be, there will always be lost electricity production. Curtailment and negative pricing shall always be around and be the last solution. Modern networks deal with it now in a smart way, by having introduced a wide range of various demand response programs, but as the grids move forward to 100% renewable energy-based systems, the need for curtailed energy will always be present. Independent power producers will need to protect their investments and have profits out of those. With the wide usage of negative prices offered, this will not happen. It will unavoidably lead to firings and job cuts (in order to maintain profitability), and this vicious circle will not end. A long-lasting investing plan on negative emissions technologies, such as DAC, will offer the opportunity to the 100% renewable energy system leading to more viable market operations. However, for this to happen, after the analysis in the US market, what it needs is to happen is to remove the existing requirements for an investor to qualify for the 45Q tax credit (meaning allowing smaller companies to receive revenue from sequestration) and establish a decade long plan, so that the future investors will not need to guess regarding future revenues. Last, in several analyses, it is stated that 45Q tax credit will offer results such as achieving DOE's R&D goals and create 0.5–3 million extra jobs [36], coupled with measures such as the carbon capture, utilization, and storage tax credits.

References

[1]   Energy Information Administration. (2018). Carbon dioxide emissions from the U.S. power sector have declined 28% since 2005. Retrieved from: https://www.eia.gov/todayinenergy/detail.php?id=37392 (Accessed 1 Aug. 2020).

[2]   Enevoldsen P. and Xydis G. Examining the trends of 35 years growth of key wind turbine components. *Energy for Sustainable Development*, 50, 18–26. DOI:10.1016/j.esd.2019.02.003.

[3]   Lazard. (2019). Lazard's levelized cost of energy analysis–Version 13.0 Lazard.

[4]   U.S. Energy Information Administration. (2020). *Levelized cost and levelized avoided cost of new generation resources*. Retrieved from https://www.eia.gov/outlooks/aeo/pdf/electricity_generation.pdf (Accessed 10 June 2020).

[5]   Apostolou D. and Xydis G. (2019). A literature review on hydrogen refuelling stations and infrastructure. current status and future prospects. *Renewable & Sustainable Energy Reviews*, 113, 109292. DOI: 10.1016/j.rser.2019.109292.

[6]   Ucal S. M. and Xydis G. (2020). Multidirectional relationship between energy resources, climate changes and sustainable development: Technoeconomic analysis. *Sustainable Cities and Society*, 60, 102210. DOI:10.1016/j.scs.2020.102210.

[7]   Panagiotidis P., Effraimis A., and Xydis G. (2019). An R-focused forecasting approach for efficient demand response strategies in autonomous micro grids. *Energy & Environment (Special Issue on Sustainable Energy Planning and Management)*, 30(1), 63–8. DOI: 10.1177/0958305X18787259.

[8]   Rasmussen N. B., Enevoldsen P., and Xydis G. (2020). Transformative multi-value business models: A bottom-up perspective on the hydrogen-based green transition for modern wind power cooperatives. *International Journal of Energy Research*. DOI:10.1002/ER.5215.

[9]   Rogelj J., Shindell D., Jiang K., *et al.* (2018). Mitigation pathways compatible with 1.5°C in the context of sustainable development. In: *Global Warming of 1.5°C. An IPCC Special Report on the impacts of global warming of 1.5°C above pre-industrial levels and related global greenhouse gas emission pathways, in the context of strengthening the global response to the threat of climate change, sustainable development, and efforts to eradicate poverty*. Retrieved from https://lup.lub.lu.se/record/d53b3238-b4c3-4363-b46c-e6c06fde199d (Accessed 10 July 2020).

[10] Lindsey R. (2020). Climate change: Atmospheric carbon dioxide. Climate.gov. Retrieved from https://www.climate.gov/news-features/understanding-climate/climate-change-atmospheric-carbon-dioxide (Accessed 28 August 2020).

[11] Realmonte G., Drouet L., Gambhir A., *et al.* (2019). An inter-model assessment of the role of direct air capture in deep mitigation pathways. *Nature Communications*, 10(3277). DOI:10.1038/s41467-019-10842-5.

[12]   Smith B. (2020). Microsoft will be carbon negative by 2030. Retrieved from https://blogs.microsoft.com/blog/2020/01/16/microsoft-will-be-carbon-negative-by-2030/ (Accessed 30 June 2020).
[13]   California Independent System Operator (CAISO). (2020). Managing oversupply. Retrieved from http://www.caiso.com/informed/Pages/Managing Oversupply.aspx (Accessed 10 July 2020).
[14]   Bird L., Cochran J., and Wang X. (2014). Wind and solar energy curtailment: Experience and practices in the United States. National Renewable Energy Laboratory. Retrieved from: https://www.nrel.gov/docs/fy14osti/ 60983.pdf (Accessed 20 June 2020).
[15]   Cheng J., Wang W., Yuan B., Zhang J., and Mi Z. (2019). Research of rational curtailment rate and development guiding mechanism of wind power in China. *IOP Conference Series: Earth and Environmental Science*, 371(5), 052059.
[16]   Zhang F., Yuan B., Zhang J. and Zhang T. (2018). Economic evaluation of power system flexibility means for improvement of wind power integration. *Journal of Global Energy Interconnection, 1*, 558–564.
[17]   Ye Q., Lu J., and Zhu M. (2018). Wind curtailment in China and lessons from the United States. Retrieved from https://www.brookings.edu/research/ wind-curtailment-in-china-and-lessons-from-the-united-states/ (Accessed 10 Aug. 2020).
[18]   Luo G., Dan E., Zhang X., and Guo Y. (2018). Why the wind curtailment of northwest China remains high. *Sustainability*, 10(2), 570.
[19]   Jorgenson J., Mai T., Brinkman G. (2017). Reducing wind curtailment through transmission expansion in a wind vision future. National Renewable Energy Laboratory. Retrieved from: https://www.nrel.gov/docs/fy17osti/ 67240.pdf (Accessed 5 Aug. 2020).
[20]   Koscis G. and Xydis G. (2019). Repair process analysis for wind turbines equipped with hydraulic pitch mechanism on the U.S. market in focus of cost optimization. *Applied Sciences*, 9, 3230. DOI:10.3390/app9163230.
[21]   Karabiber O.A. and Xydis G. (2019). Electricity price forecasting in Danish day-ahead market using TBATS, ANN and ARIMA methods. *Energies (Special Issue 'Demand Response in Electricity Markets')*, 12(5), 928. DOI:10.3390/en120509282019.
[22]   Karabiber O.A. and Xydis G. (2021). A review of the day-ahead natural gas consumption in Denmark. Starting point towards forecasting accuracy improvement. *International Journal of Coal Science and Technology*, 8, 1–22. DOI: 10.1007/s40789-020-00331-2.
[23]   Karabiber O.A. and Xydis G. (2020). Forecasting day-ahead natural gas demand in Denmark. *Journal of Natural Gas Science & Engineering*. DOI:10.1016/j.jngse.2020.103193.
[24]   Strommer L. and Conant S.(2018). Conservation value of koa (*Acacia koa*) reforestation areas on Hawaii Island. *Pacific Conservation Biology*, 24(1), 35–43.

[25]   Vakalis S., Moustakas K., Sénéchal U., *et al.* (2017). Assessment of potassium concentration in biochar before and after the after-burner of a biomass gasifier. *Chemical Engineering Transactions*, 56, 631–636.

[26]   Kulyk N. (2012). Cost-benefit analysis of the biochar application in the US cereal crop cultivation [Internet]. University of Massachusetts–Amherst. School of Public Policy Capstones.

[27]   Low S. and Schäfer S. (2019). Tools of the trade: Practices and politics of researching the future in climate engineering. *Sustainability Science*, 14(4), 953–962.

[28]   Fasihi M., Efimova O., and Breyer C. (2019). Techno-economic assessment of CO_2 direct air capture plants. *Journal of Cleaner Production*, 224, 957–980. DOI:10.1016/j.jclepro.2019.03.086.

[29]   Ranjan M. and Herzog H.J. (2011). Feasibility of air capture. *Energy Procedia*, 4, 2869–2876. DOI:10.1016/j.egypro.2011.02.193.

[30]   Jacobson M.Z. (2020). Why not synthetic direct air carbon capture and storage (SDACCS) as part of a 100% wind-water-solar (WWS) and storage solution to global warming, air pollution, and energy security. In: Jacobson, M.Z., 100% Clean, Renewable Energy and Storage for Everything. New York, NY: Cambridge University Press.

[31]   Internal Revenue Service. (2019). Fossil energy internal revenue code tax fact sheet IRS. Retrieved from https://www.irs.gov/newsroom/fact-sheets (Assessed date: 28 July 2020).

[32]   Roungkvist J.S., Enevoldsen, P., and Xydis G. (2020). High-resolution electricity spot price forecast for the Danish power market. *Sustainability, SI: Sustainable and Renewable Energy Systems, Sustainability*, 12, 4267. DOI:10.3390/su12104267.

[33]   Apostolou D., Enevoldsen P., and Xydis G. (2019). Supporting green urban mobility – The case of a small-scale autonomous hydrogen refuelling station. *International Journal of Hydrogen Energy (Special Issue of SI:ICH2P-2018)*, 44 (20), 9675–9689. DOI: 10.1016/j.ijhydene.2018.11.197.

[34]   Bergstrom J.C. and Ty D. (2017). Economics of carbon capture and storage. *IntechOpen*. DOI:10.5772/67000.

[35]   Anderson C. (2019). Decrement carbon: 'Stripe's negative emissions commitment. Retrieved from https://stripe.com/blog/negative-emissions-commitment (Accessed 10 July 2020).

[36]   U.S. Department of Energy, Internal Revenue Code Tax Fact Sheet. (2019). Retrieved from https://www.energy.gov/sites/prod/files/2019/10/f67/Internal%20Revenue%20Code%20Tax%20Fact%20Sheet.pdf (Accessed 10 Sept. 2020).

Chapter 11

Exergy analysis of a small-scale trigenerative compressed air energy storage system

Raghuveera Sai Sarath Dittakavi[1], David S-K. Ting[1], Rupp Carriveau[1] and Mehdi Ebrahimi[1]

Trigenerative compressed air energy storage (T-CAES) capitalizes on the heat of the compression process, something that is often wasted in more conventional compressed air energy storage (CAES) approaches. A T-CAES system with a 4 kW compressor and 2 kW turbine is thermodynamically analyzed in this study. Exergy analyses performed on each component in the system identify specific areas for improvement. It is found that, under actual conditions, more than half of the total exergy destruction is caused by the accumulator and about a quarter of the destruction is caused by the pressure regulator and turbine. Further, the pressure regulator, accumulator, and turbine offer 66%, 27%, and 32% of individual component recoverable exergies, respectively. These recoveries can improve the overall exergy efficiency of the system by 35%.

Keywords: Trigeneration; exergy; unavoidable; relative exergy; small-scale; destruction ratio

11.1 Introduction

Replacing conventional energy conversion systems which run on fossil fuels with renewable energies is an effective way to help mitigate economic and environmental concerns. The intermittent nature of many renewable energy resources and the energy supply–demand mismatch call for energy storage. There are several energy storage systems, such as pumped hydroelectric storage systems, CAES systems, thermal energy storage (TES), batteries, and supercapacitors, among others. Among these options, CAES has great potential due to its high reliability, low capital and maintenance costs, and good part-load performance [1]. In CAES

[1]Turbulence and Energy Laboratory, Ed Lumley Centre for Engineering Innovation, University of Windsor, Windsor, Canada

systems, the energy to be stored is used to compress ambient air into storage tanks. When the need arises, the compressed air is converted to electrical energy by expanders [1–3]. The process of compressing air generates heat, and how that heat is dealt with is the main criterion in the classification of CAES systems, as shown in Figure 11.1. There are three main types of CAES systems: (1) diabatic CAES (D-CAES), (2) adiabatic CAES (A-CAES), and (3) isothermal CAES (I-CAES).

In D-CAES, the heat generated in the compression process is lost to the environment as waste heat [5–7]. Hence, an external heat source is needed to prevent condensation on the expander, which compromises system efficiency, which is around 40%–53% [8]. This drawback has been rectified to an extent with the introduction of A-CAES systems. In A-CAES, the heat produced during the compression is collected and stored in a TES system, and then used to preheat the air before expansion. Because of the recuperation and use of heat generated by compression, the system efficiency of A-CAES can be close to 65% [9]. This technology is significantly more advanced than D-CAES, and, thus, is also known as "advanced CAES."

In the I-CAES process, the temperature is kept stable during the compression process. This lowers the power required to run the compressor below the amount required to run an adiabatic compressor with the same pressure ratio. During the expansion, the associated heat is supplied constantly to ensure expansion at a constant temperature. Despite this innovation, near-isothermal compression is yet to be used industrially [10]. The CAES approaches differ widely based on quantitative parameters like energy density, start-up time, and cycle efficiency, as well as in more qualitative measures, like their states of commercialization.

The concept of trigenerative compressed air storage (T-CAES) [11,12] derives from the adiabatic classification. The (T-CAES) system simplifies the actual A-CAES by removing the regenerative air reheating. The heat eliminated during the compression phase is stored and, rather than being utilized to reheat the air at the turbine inlet, may be used to deliver a thermal energy demand [11]. As a corollary, air enters the expansion train at a low temperature, and chilling energy is obtained at the expander's outlet, without the need of an inverse cycle. Placing such a plant close to the energy user facilitates the effective utilization of all the energy streams.

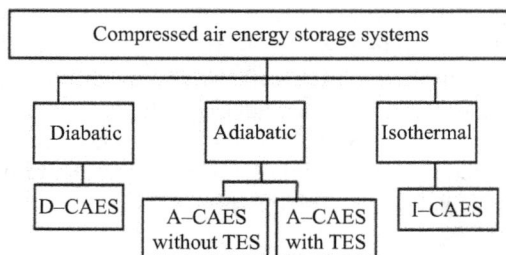

Figure 11.1   Classification of CAES based on the type of design [4]

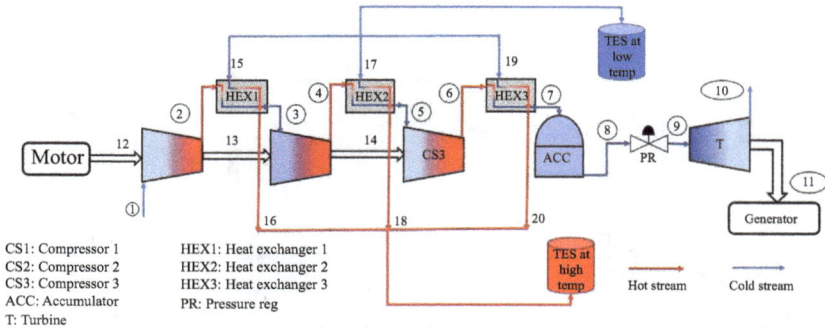

Figure 11.2 Example of T-CAES system [13]

The schematic of T-CAES system is shown in Figure 11.2. A CAES system generally consists of three phases. First is the compression phase in which compressors and heat exchangers are the key components. When mechanical or electrical energy is available, the compressor train begins pulling air from the environment into high-pressure reservoirs. In between each compression phase, heat exchangers are used to capture the heat generated and store them in the TES tanks. Second is storage phase where the compressed air with high pressures is stored in the storage tanks. Finally, the system ends with expansion phase where power is generated through expansion. Since thermal energy is stored and utilized on a daily basis, the heat storage does not pose any particular technological problems, as it may be simply achieved through a properly insulated tank. There is a significant increase in system efficiency to 68%, which is slightly higher than the A-CAES system's efficiency of 63% [9]. When mechanical energy is necessary from the T-CAES, the air stored in the high-pressure storage tanks is expanded through a reheated multistage turbine train, without usage of additional fuel. As for the compression, the optimal choice for the expansion technology depends on the expansion ratio and mass flow rate; it is commonly assumed that one could switch to a small centripetal turbine whilst still having to use a volumetric compressor to obtain high efficiency.

A D-CAES system along with its *T–S* plot diagram is illustrated in Figure 11.3. As you can see, isentropic compression takes place in process 1-2. This is because we are assuming ideal working conditions. The storage is considered as isobaric/adiabatic so no change in the parameters from the outlet of the storage tank. When the power is needed, discharging takes place. Isobaric heat addition takes place in process 2-3 as the air from the storage tank is preheated prior to expansion. Process 3-4 indicates isentropic expansion, where output air is supplied to generator to generate electricity.

An A-CAES is illustrated in Figure 11.4(a) and the *T–S* plot for that system is shown in Figure 11.4(b). Unlike D-CAES systems, the heat lost by the compressor during the compression is recovered through heat exchangers and then stored in the

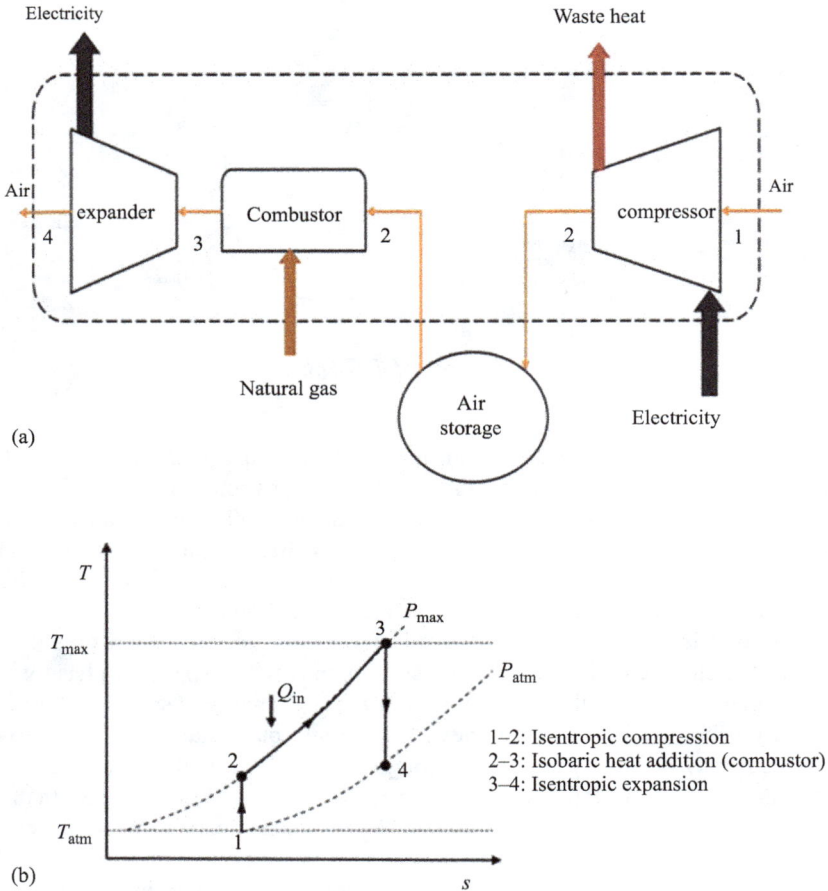

Figure 11.3   D-CAES system: (a) schematic [9]; (b) T–S diagram

TES systems. This heat is reused for preheating the air before expansion. This way it reduces the use of natural gas and also reduces the emissions. Since a heat exchanger is used after the compression to capture the heat lost, there will be slight drop in the temperature, which is shown in the *T–S* diagram as process 2-3. Process 3-4 indicates the isobaric heat addition and 4-5 indicates the isentropic expansion.

The I-CAES system and its *T–S* plot diagram are illustrated in Figure 11.5. Main difference between I-CAES and A-CAES is the heat transfer of the air. In case of A-CAES, the temperature of the air increases significantly during air compression. High-temperature air exchanges heat with thermal stores outside compressors. In case of ICAES, the heat transfer takes place inside compressors. This allows the temperature of the air to be close to ambient during air compression, illustrated as process 1-2 in Figure 1.5(b), indicating isothermal compression.

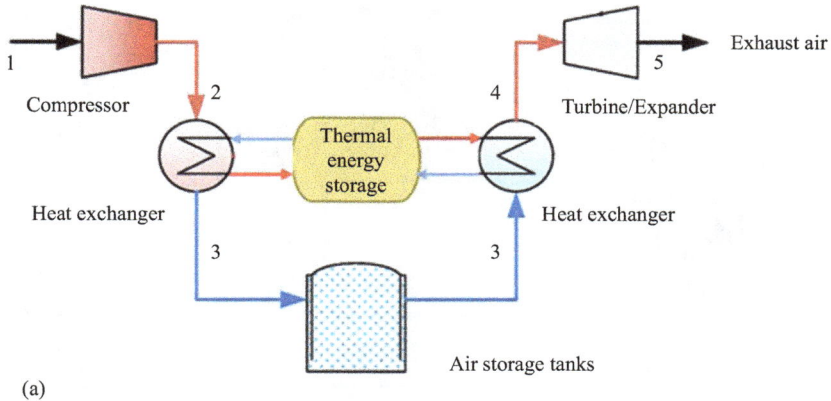

(a)

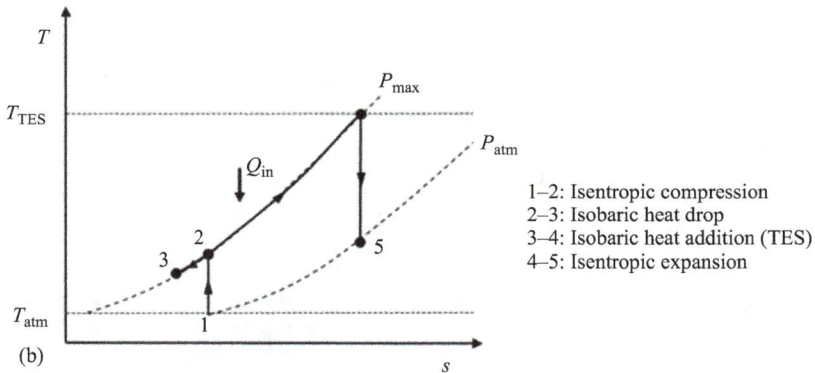

1–2: Isentropic compression
2–3: Isobaric heat drop
3–4: Isobaric heat addition (TES)
4–5: Isentropic expansion

(b)

Figure 11.4 A-CAES system: (a) schematic [3]; (b) T–S diagram

The heat generated from the isothermal compression is reduced. The energy loss related to the heat transfer in thermal stores decreases. This eventually results in increased compression efficiency and reduced heat losses to the environment. Processes 2-3 and 3-4 indicate isobaric heat addition and isentropic expansion, respectively.

Researchers have studied various configurations of isothermal systems. These include systems with near-isothermal compression and expansion process, and systems with isothermal compression and isentropic expansion. We can achieve more specific work output through isothermal expansion than isentropic expansion. In real life, however, there are many challenges hindering the full realization of these ideal conditions. For this reason, we have provided a one-of-a-kind study that has experimentally accomplished isothermal compression via appropriate heat transfer in the compressor [14]. This is illustrated in Figure 11.5(a), where the expansion process is isentropic. This configuration has resulted in increased

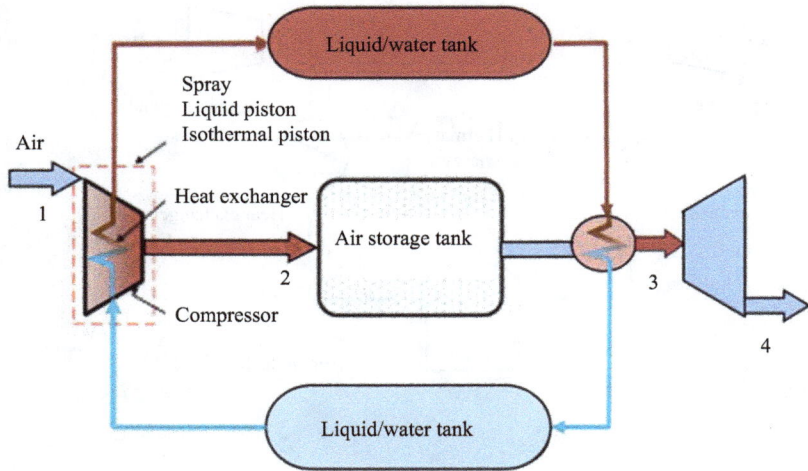

(a)

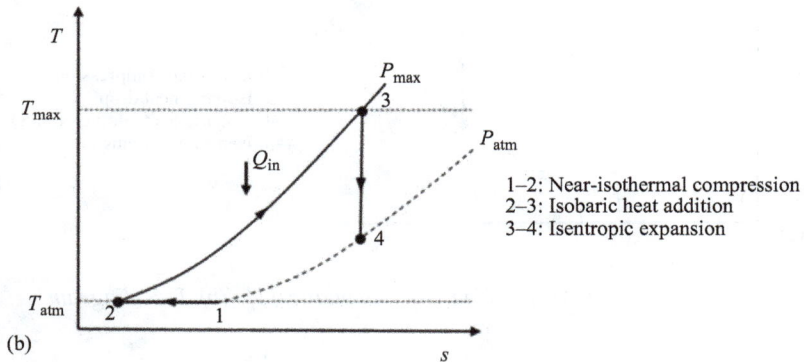

(b)

Figure 11.5   I-CAES system: (a) schematic [14]; (b) T–S diagram

compression efficiency from 11% to 28% [14], which ultimately resulted in significant improvement in overall system performance. The manufacturing of the first prototype is currently underway.

11.1.1   Literature review

The flowchart shown in Figure 11.6 summarizes the literature review. CAES research and advancement has been very active in recent years, with various aboveground and underground pilot plants being tested all over the world. Thermodynamic analysis was performed for a CAES combined cycle (CAES-CC) by Liu *et al.* [15]. Zhao *et al.* [16] studied a system consisting of a CAES system, and a Kalina cycle to recover waste

```
┌───────────────────────────────────────────────────────────────┐
│                Compressed air energy storage systems            │
└───────────────────────────────────────────────────────────────┘
                                │
┌───────────────────────────────────────────────────────────────┐
│                            D-CAES                               │
│             Thermodynamic assessment on CAES [15]               │
│           Study on CAES combined with Kalina cycle [16]         │
└───────────────────────────────────────────────────────────────┘
                                │
┌───────────────────────────────────────────────────────────────┐
│                            A-CAES                               │
│          A-CAES with focus on the effect of TES [4, 17–19]      │
│   Effect of individual components on the A-CAES system          │
│                         performance [8, 20, 21]                 │
│     Innovative works on A-CAES, e.g., underwater A-CAES, etc.    │
│                              [6, 22–28]                          │
└───────────────────────────────────────────────────────────────┘
                                │
┌───────────────────────────────────────────────────────────────┐
│                            T-CAES                               │
│               Thermodynamic assessment [29–31]                  │
│   Experimental studies on T-CAES, e.g., small scale systems     │
│                       [11, 13, 32–34]                           │
└───────────────────────────────────────────────────────────────┘
                                │
┌───────────────────────────────────────────────────────────────┐
│                             GAP                                 │
│    Lack of exergy analyses on small-scale T-CAES system and     │
│   study on effect of individual components over the performance │
│                         of the system                           │
└───────────────────────────────────────────────────────────────┘
```

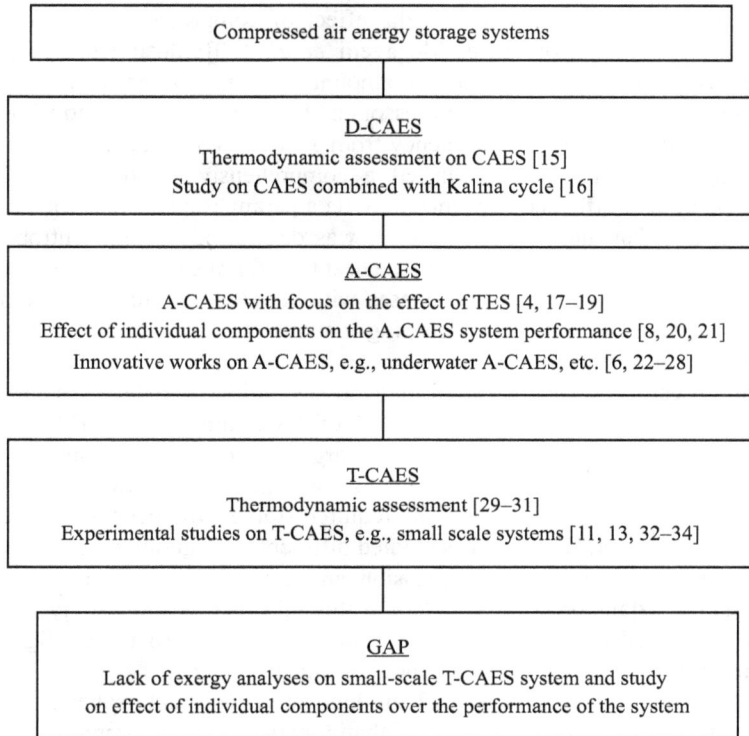

Figure 11.6   A summary of literature review on CAES

heat was presented. The system had an efficiency higher by 4% compared to a stand-alone, regenerative CAES system. The overall efficiency of the system was about 10% higher than the conventional, non-regenerative reference CAES. The world's first A-CAES plant was built by HYDROSTOR in Goderich, Ontario, Canada. It stands as the first utility-scale plant with a 1.75 MW power output and a 10+ MWh storage capacity. Budt *et al.* [4] conducted a comprehensive literature review on CAES systems, and classified A-CAES based on the temperature level of the TES. Budt and Wolf [17] proved that a low-temperature TES (below 200 °C) maintained a high level of round-trip efficiency, as well as surmounting technological problems associated with the high-temperature outputs of compressors. Zhang *et al.* [18] examined the effect of TES on the efficiency of A-CAES, finding that a quantity of heat can be left in the TES which could be used to further improve the efficiency of the system. In response, Zhou *et al.* [19] analyzed the effect recovering the exhaust heat released from the output of the last stage turbine had on the system efficiency of A-CAES and found efficiencies approaching 68.7%.

Recent research has focused on the effect of compressor and turbine efficiencies on system performance. Mozayeni *et al.* [20] illustrated that storage pressure has a substantial effect on the amount of energy stored, concluding that increasing the efficiency of the compressors and turbines from 65% to 95% could increase the round-trip electric efficiency from 35% to 74%. In agreement with this conclusion, Luo *et al.* [8] established a comprehensive model for A-CAES, focusing on system efficiency optimization via a parametric analysis; the principal conclusion was that the system efficiency was dominated by the isentropic efficiency of the turbine, compressors, and the heat transfer rate of the heat exchangers. He *et al.* [21] studied compression phases with variable pressure ratios and optimized the compression efficiency, keeping it above 80% by varying the rotational speed and the blade inlet angle.

Many researchers have proposed innovative solutions to reduce system losses. Houssainy *et al.* [22] proposed a hybrid of high-temperature TES and low-temperature A-CAES that included a turbocharger unit that supplements mass flow rate alongside the air storage. Their results show that the addition of the turbocharger has the potential to mitigate the required storage volume and the pressure, thus reducing the cost. Kim [23] has studied different configurations of CAES with adiabatic or quasi-isothermal compression and expansions, as well as constant volume and constant pressure air storage through energy and exergy analyses. Outcomes revealed that constant pressure and isothermal process configurations performed best of the configurations they examined.

Mazloum *et al.* [24] introduced an innovative constant isobaric A-CAES approach that included multistage adiabatic expansion and compression that resulted in a round-trip electrical efficiency of 54%. Bagdanavicius and Jenkins [25] performed exergy analyses of a CAES system combined with hot water TES; results revealed a 75% energy efficiency. A pilot project of underground A-CAES built in Switzerland [26,27] was studied, showing a 63% round-trip efficiency. Ebrahimi *et al.* [6] performed traditional and advanced exergy analyses on underwater CAES systems. Their results highlighted that 76% of destroyed exergy was avoidable, emphasizing the significant potential for improvement of the system. Transient thermodynamic modeling of an underwater CAES system was also conducted by Carriveau *et al.* [28], demonstrating the significance of considering transients for the characterization and potential improvement of CAES performance.

CAES also enables the cogeneration of heat and cooling demands, which promotes the concept of T-CAES. Many configurations have been proposed which differ depending on the manner in which the compression's heat is used. Some researchers have devoted the heat produced during the charge phase for heating purposes during the discharge phase, when the electricity and the cooling energy are produced. Arabkooshar *et al.* [29] applied this concept to a 300 MW wind farm, proving the potential of their recommended configuration to support district heating and cooling networks. The values of power-to-power, power-to-heat, and power-to-cooling efficiencies of this system were 30.6%, 92.4%, and 32.3%, respectively.

Lv *et al.* [30] employed a thermodynamic model to assess the monthly economic and energy performance of T-CAES used for electrical energy peak load shifting at a hotel. The results showed that the trigeneration system worked efficiently at comparatively low pressures, and the efficiency was able to reach 76.3% at 15 bars. Liu *et al.* [31] instituted a configuration of T-CAES and focused on the discharge process formed by a scroll expander. They evaluated its polytropic exponent as a function of the ambient temperature and examined the effects that maximum storage pressure and the expansion ratio had on the system's performance.

Additionally, there were also studies of configurations that can produce both cooling and heating energies utilizing the heat stored during the expansion process. Han and Guo [32] developed a configuration from A-CAES that enabled the liberation of cooling energy from the last stage of expansion and the delivery of excess heat as heating energy. A variable expansion ratio was proposed to increase the electric efficiency, which reached 44.5%. Li *et al.* [11] introduced a new tri-regenerative system to meet the end user cooling, heating, and electricity demands of a small 52 kW office building in Chicago; a global storage electric efficiency of 50% was achieved. A T-CAES system for a small-scale, stand-alone photovoltaic power plant with 3.7 kW electric compressor input and 1.7 kW expander electric output was proposed by Jannelli *et al.* [33]. The system fulfilled electric energy and cooling demand for a radio base station with a coefficient of performance of 0.62 and an electric efficiency of 57%.

Venkataramani *et al.* [34] constructed an experimental T-CAES setup with a wind turbine (3.2 kW), a scroll compressor, expander, and a reservoir (with a capacity of 400 L and maximum pressure of 8 bars). Increasing discharge mass flow rate increased round-trip efficiency to a maximum of 22% at the maximum flow rate. Cheayb *et al.* [13] performed a thermodynamic assessment of a small-scale T-CAES system. They demonstrated that the Joule–Thomson effect leads to a temperature change across the pressure regulator and that the supposition of constant temperature is no longer true as stated in previous models of CAES systems. Their work also represented the first reliable model based on experimental data for small-scale CAES, which will be useful for future trigenerative system studies.

Most of the literature cited above related to CAES, with focuses on theoretical modeling, system configurations, and parameter optimization of T-CAES, and no exergy analyses have been performed on small-scale T-CAES. Further, there has been little focus on the effect that individual components have on the whole system performance. Thus, this current work could aim to fill this gap by performing exergy analysis over the components of small-scale (4 kW compressor and 2 kW turbine) T-CAES which could help us in finding out the components with highest exergy destruction rate and then offering subsequent improvement recommendations.

11.2 Thermodynamic laws

A description of any thermodynamic system employs the laws of thermodynamics that form an axiomatic basis. The first law specifies that energy can be exchanged

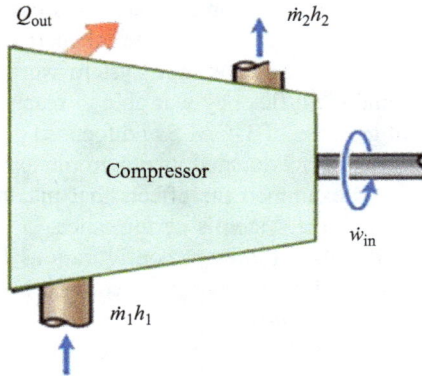

Figure 11.7   Steady-state energy flow of a compressor

between physical systems as heat and work. The second law defines the existence of a quantity called entropy that describes the direction, thermodynamically, that a system can evolve, as well as quantifying a system's state of order. Entropy can also be used to quantify the useful work that can be extracted from the system. In this chapter, we will discuss the first and second laws of thermodynamics and how they can be employed in our CAES system.

11.2.1   The first law of thermodynamics

The first law states that *energy can neither be created nor destroyed; it can be transformed from one form to another.* Under steady-state operation, the rate of energy entering a system is equal to that leaving the system, i.e.,

Energy in − Energy out = Energy accumulation   (11.1)

For example, a simple compressor is shown in Figure 11.7 with a single inlet and outlet.

Therefore, the rate of mass entering the compressor is equal to that leaving the system, i.e., $\dot{m}_1 = \dot{m}_2 = \dot{m}$.

Additionally, heat is lost from the system and work is supplied to the system. Under steady-state operation, the rate of energy entering the system is equal to that exiting the system, i.e.,

$$\dot{E}_{in} = \dot{E}_{out} \qquad (11.2)$$

$$\dot{W}_{in} + \dot{m}_1 h_1 = Q_{out} + \dot{m}_2 h_2 \qquad (11.3)$$

Here,

$\dot{E}_{in}, \dot{E}_{out}$ = rate of net energy transfer by heat, work, and mass

$\dot{W}_{in}$ = rate of work input
$\dot{Q}_{out}$ = rate of heat transfer from the system
$\dot{m}_1, \dot{m}_2$ = mass flow rate in the system at input and output.

Energy analysis can be explained by considering the above example and Figure 11.7 for the energy flow. For our understanding, input and output parameters drawn from Cheayb *et al.* [13] are shown in Table 11.1. These values are required to find out the enthalpies of the working fluid, i.e., air using (11.3) as needed in the energy analysis to determine the parameters like work input, etc.

11.2.2   Exergy

Exergy is consumed due to irreversibility, and exergy consumption is proportional to entropy creation. The main important difference between energy and exergy is that energy is conserved, while exergy, a measure of energy quality or work potential, can be consumed. The general exergy balance equation for any kind of process is expressed as shown in (11.4):

$$\text{Exergy}_{in} - \text{Exergy}_{out} - \text{Exergy consumed} = \text{Exergy accumulation} \quad (11.4)$$

Exergy analysis is a very convenient method to assess the performance of energy conversion systems. It helps us to determine how a source can be used effectively [15]. It is also called the second law of thermodynamic analysis.

We considered a generalized one-dimensional wall with steady-state heat transfer, as shown in Figure 11.8, to explain the concept of exergy. To determine the total exergy destruction through this heat transfer process, we consider the system, including the regions on both sides of the wall that experience the temperature change. One side of the system boundary becomes the room temperature T_1, while the other side is the temperature of the outdoors T_2, as shown in the figure. This is a closed system, so there is no mass transfer and only heat is transferred from one side to another side of the wall.

According to the exergy balance analysis, the amount of exergy entering the system must be equal to the amount of exergy leaving, plus the amount of exergy destroyed or consumed, as shown in (11.5). On applying this concept to the above-considered steady-state heat-transfer example, the rate of exergy destroyed is deduced as (11.6):

Table 11.1   Table providing the parametric data of the compressor

Parameter	Value
Mass flow rate (kg/s)	0.0039
Input pressure (bar)	1.25
Output pressure (bar)	7.7
Input temperature (°C)	23.5
Output temperature (°C)	118

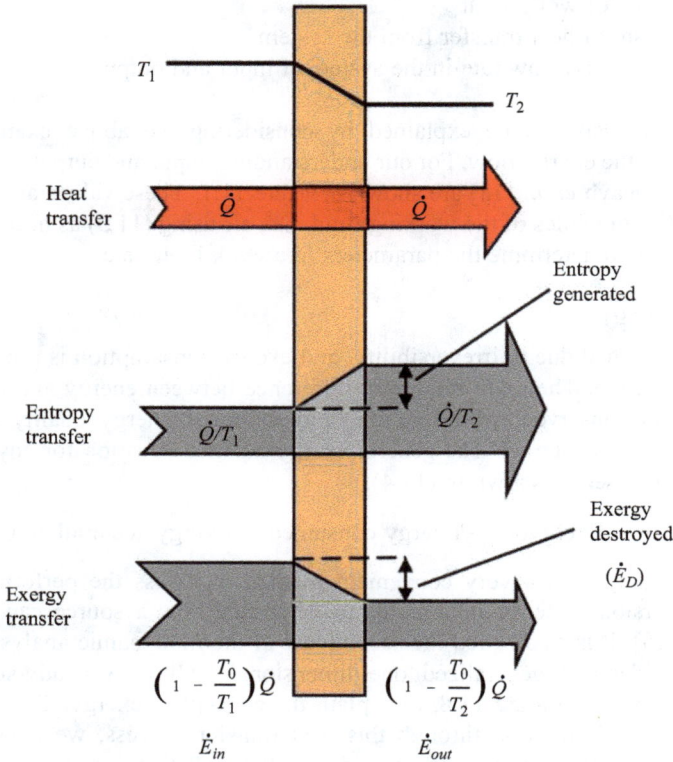

Figure 11.8   Exergy flow of steady-state heat transfer through a wall

$$\dot{E}_{\text{in}} - \dot{E}_{\text{out}} = \dot{E}_D \tag{11.5}$$

$$\dot{E}_D = \dot{Q}\left(1 - \frac{T_0}{T_1}\right)_{\text{in}} - \dot{Q}\left(1 - \frac{T_0}{T_2}\right)_{\text{out}} \tag{11.6}$$

where

$$\dot{E}_{\text{in}}, \dot{E}_{\text{out}} = \text{rate of exergy transfer by heat} = \dot{Q}\left(1 - \frac{T_0}{T}\right)$$

$\dot{E}_D$ = rate of exergy destructed
$\dot{Q}$ = rate of heat transfer
T_0 = ambient temperature (K).

11.2.3   Conditions used in exergy analysis

Terms like *actual conditions, unavoidable conditions, relative exergy destruction,* and *unavoidable exergy* which were used in the current work are discussed in this section. Also, numerical examples are given to improve understanding.

The **actual conditions** of the system are the operating conditions of the system. Here, actual efficiencies, i.e., the conditions those are operating in real life, are considered.

The **unavoidable working conditions** of a system are the parameters that are determined by considering the conditions under the assumption that each component operates with unavoidable thermodynamic inefficiency [16]. In more general terms, these are conditions that just cannot be predicted or avoided in the foreseeable future. The unavoidable part in the exergy destruction represents the part that cannot be eliminated even with the technological advancements available today. The advanced analysis is performed by analyzing each component separately as if the component is removed from the system. The conditions or the assumption for the advanced analysis is made considering future enhancements that can be made for the component [35]. For this purpose, decision makers must understand the working of the entire system and rely on conditions that can improve the system. In general, it can be said that the unavoidable conditions that are considered are better than the real working conditions but are not the ideal theoretical working conditions for the component.

An explanation for both working conditions is provided by considering an example of the heat exchanger that was used in our analysis. First, the input and output parameters required by the heat exchanger under actual working conditions are shown in Table 11.2 [13]. The table displays the parameters of both working and cooling mediums. The effectiveness of the heat exchangers under actual conditions was taken as 0.583 [13].

In the similar way, the input parameters and output parameters of the heat exchanger under unavoidable conditions are given in Table 11.3. These are the values

Table 11.2   Input and output parameters of a heat exchanger under actual conditions

Component	Parameter	Working fluid (air)		Cooling medium	
		In	Out	In	Out
HEX	T (K)	391	335	295	334.3
	P (bar)	7.7	6.9	1.013	1.013
	$\dot{m}$ (kg/s)	0.0039	0.0039	0.0043	0.0055

Table 11.3   Input and output parameters of a heat exchanger under unavoidable working conditions

Component	Parameter	Working fluid (air)		Cooling medium	
		In	Out	In	Out
HEX	T (K)	386.2	304.1	295	359.7
	P (bar)	8.1	7.6	1.013	1.0.13
	$\dot{m}$ (kg/s)	0.0035	0.0035	0.0044	0.0044

those were determined by considering the highest efficiency of the component; here in this case, it is the effectiveness which is considered as 0.9 [35,36].

Now, using the equations given, we can calculate the exergy of fuel and product, as well as the amount of exergy that was destroyed. We consider the fuel and production method to calculate exergy destroyed. The working fluid is considered to be air and the cooling medium is water. As per this method, the amount of exergy destroyed would be as shown through (11.7)–(11.9) [13].

$$\dot{E}_D = \dot{E}_F - \dot{E}_P \tag{11.7}$$

$$\dot{E}_F = \dot{m}_{c,a} \left[c_{p,a} \cdot \left(T_{in,a} - T_{out,a} - T_{amb} \cdot \log \frac{T_{in,a}}{T_{out,a}} \right) \right]$$

$$+ T_{amb} \cdot R_g \cdot \log \frac{P_{in,a}}{P_{out,a}} \tag{11.8}$$

$$\dot{E}_P = \dot{m}_{c,w} \left[c_{p,w} \cdot \left(T_{out,w} - T_{in,w} - T_{amb} \cdot \log \frac{T_{out,w}}{T_{in,a}} \right) \right] \tag{11.9}$$

where

$\dot{m}_{c,a}$ = mass flow rate of the working fluid (air) (kg/s)
$\dot{m}_{c,w}$ = mass flow rate of the cooling fluid (water) (kg/s)
$c_{p,a}$ = specific heat of the working fluid
$c_{p,w}$ = specific heat of the cooling fluid
$\dot{E}_F$ = exergy of fuel
$\dot{E}_P$ = exergy of product
R_g = gas constant (N m/kmol K)
$T_{in,a}, T_{out,a}$ = input and output temperatures of the working fluid (air) (K)
$T_{out,w}, T_{in,w}$ = input and output temperatures of the cooling fluid (water) (K)
T_{amb} = ambient temperature (K)
$P_{in,a}, P_{out,a}$ = input and output pressures of the working fluid (air).

The exergy flow in the heat exchanger is shown in Figure 11.9. By using the input and output parameters provided in Tables 11.2 and 11.3 and Equations (11.7)–(11.9), the amount of exergy destroyed in the heat exchanger under actual conditions and unavoidable conditions are given in Table 11.4. With these results, we can understand how much of energy given to the system is actually utilized by the component and how much has been wasted. On the other hand, the results under unavoidable conditions help us understand how far we can improve the system by knowing the amount of exergy loss that is inevitable and the exergy loss that can be avoidable.

A better understanding of the components can be achieved by comparing the system's performance under one condition with the performance under slightly altered conditions. Through this process, we can better see which components are more sensitive and effective for the efficiency improvement. The relative exergy destruction of a component can be determined as the ratio of difference between exergies destroyed in two state conditions to the average of those destruction rates which can be written as (11.10):

Cooling fluid

$\dot{E}_{in}$

Working fluid

$\dot{E}_{in}$

Working fluid

$\dot{E}_{out}$

$$\dot{E}_D = (\dot{E}_{in} - \dot{E}_{out}) - (\dot{E}_{in} - E_{out})$$

Cooling fluid   Working fluid   Cooling fluid

$\dot{E}_{out}$

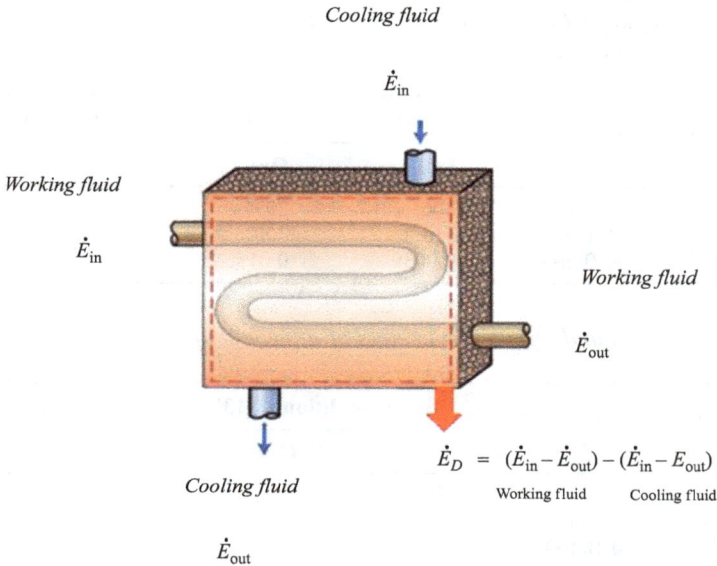

Figure 11.9   Exergy flow in heat exchanger

Table 11.4   Exergy values of the heat exchanger under actual and unavoidable conditions

Component	Actual conditions			Unavoidable conditions		
	$E_{F,\ K}$ (W)	$E_{P,K}$ (W)	E_{D_UN} (W)	$E_{F,\ K}$ (W)	$E_{P,K}$ (W)	E_{D_UN} (W)
Heat exchanger	41	13	27	58	38	20

$$E_{D\_rel} = \frac{E_{D,K2} - E_{D,K1}}{0.5 \cdot (E_{D,K2} + E_{D,K1})} \tag{11.10}$$

Here,

E_{D_rel} = relative exergy destruction

$E_{D,K1}$, $E_{D,K2}$ = exergy destruction at State conditions 1 and 2.

Let us consider two components—i.e., accumulator and the turbine—from our CAES system to understand the relative exergy concept. The accumulator's parameters under the first and second state conditions are illustrated in Table 11.5. Input temperature of the component has been varied by 3 °C from the first state condition to the second state condition. Based on this change, the other parameters of the component were determined. In a similar way, the two state conditions for the turbine are illustrated in Table 11.6. Upon using the exergy equations as discussed above and substituting the corresponding parameters, the relative exergies of the two components were determined and are shown in Table 11.7.

Table 11.5   Input and output parameters of an accumulator under two state conditions

Component	Parameter	State condition 1 [13]		State condition 2	
		In	Out	In	Out
Accumulator	T (K)	306	297	309	300
	P (bar)	313.8	300	496.3	473.5
	$\dot{m}$ (kg/s)	0.0039	0.015	0.0043	0.016

Table 11.6   Input and output parameters of a turbine under two state conditions

Component	Parameter	State condition 1 [13]		State condition 2	
		In	Out	In	Out
Turbine	T (K)	283	281.25	286	284.5
	P (bar)	5	1.013	9.48	1.5
	$\dot{m}$ (kg/s)	0.0136	0.0136	0.0144	0.0144

Table 11.7   The results of relative exergy destruction

Component	$E_{F,\,K}$ (kW)	$E_{P,K}$ (kW)	E_D (kW)	E_{D_rel} (kW)
Accumulator	81	78.2	2.87	0.02
Turbine	1.86	0.538	1.3	0.51

In this way, we can determine the relative exergy destruction. As mentioned, the two components were exposed to the same amount of change, i.e., 3 °C increase in the input temperature. Upon this variation imposed on the two components, one of them displayed higher relative destruction and the other one has lesser destruction. From this result, we can conclude that it is not necessarily the component with highest destruction rate that exhibits more sensitivity. Even components with lower destruction could exhibit higher sensitivity to the changes applied.

11.3   T-CAES system setup and analysis

This chapter provides a description of the system considered for these analyses and discusses the methodology of the analyses.

11.3.1   System description

Generally, CAES has three phases: (1) charge or compression phase, (2) storage phase, and (3) the discharge or expansion process. Our system analyses are based

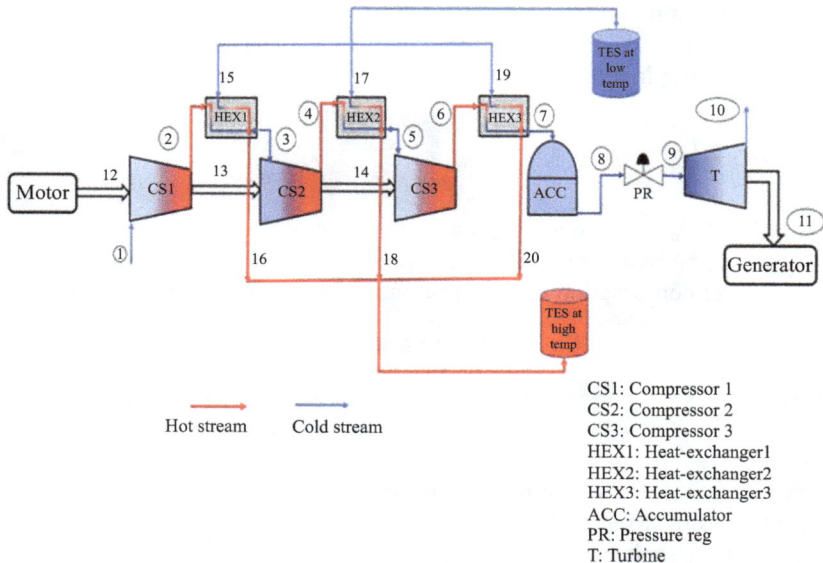

Figure 11.10   Schematic of the T-CAES system under consideration

on the previous experimental work of Cheayb *et al.* [13]. Figure 11.10 illustrates the schematic of the system with the charging, storing, and discharge phases.

The six main components of T-CAES are compressors and heat exchangers for the charging phase, a storage tank for the storage phase, a pressure regulation valve, an air motor (expander) for discharge phase, and, lastly, TES. In this section, conventional exergy analyses of these components are presented, with the exception of the TES.

Here, only steady-state charge and discharge phases are considered. This model is employed to evaluate both the performance of potential T-CAES configurations and prospective recommendations for improvement. The following sections discuss the thermodynamic analyses and results. The following assumptions were made to reduce modeling complexity:

1. Pressure losses at the admission, through the discharge valve, and in the heat exchangers are not considered.
2. Compressed air is considered as a perfect gas, except in the regulating valve.
3. A constant global compression ratio is considered.
4. Kinetic and potential energies are neglected.

11.3.2   Compressors

For limited mass flow rates and high-pressure ratios, volumetric compressors are suitable [37]. Multistage compression is required to lower the specific energy consumption for air mass storage. Inherently, volumetric compressors lose some amount of heat to the environment, which can be described by a polytropic coefficient $\eta_c < \gamma$. Thus, the output temperature of each stage can be calculated multiplying temperature in with the compression ratio as shown in (11.11):

$$T_{\text{out},c,i} = T_{\text{in},c,i} \cdot \beta_{c,i}^{\frac{n_{c,i}-1}{n_{c,i}}} \tag{11.11}$$

where

$T_{\text{out},c,i}$ and $T_{\text{in},c,i}$ = input and output temperatures of the compressor
$\beta_{c,i}$ = compression ratio of the compressor.

The power consumption of the stage one compressor can be determined using (11.12):

$$\dot{W}_{\text{th},c,i} = \frac{\gamma - 1}{\gamma} \frac{n_{c,i}}{n_{c,i} - 1} \dot{m} T_{\text{in},c,i} C_{p,a} \left[\beta_{c,i}^{\frac{n_{c,i}-1}{n_{c,i}}} - 1 \right] \tag{11.12}$$

where

$\dot{W}_{\text{th},c,i}$ = power input to the compressor
$\dot{m}$ = mass flow rate in the compressor
$n_{c,i}$ = polytropic coefficient
γ = specific heats ratio or heat capacity index or adiabatic index.

Gamma (γ) is derived from the relation of specific heats and the gas constant R, i.e., $C_p - C_v = R$. Input temperature of each stage hinges on both the previous heat exchanger and previous stage compressor's output temperature. The pressure ratio of the three compressors is constant. It should be noted that, for a fixed pressure ratio, the optimal distribution is symmetric, which is different from a manufactured compressor [37]. Air humidity is ignored because the compressor is equipped with dehumidifier.

11.3.3   Heat exchangers

Countercurrent air-to-air heat exchangers have been selected for this system. The thermal energy produced from compression is transferred from the heat source (compressed air) to the heat sink, which is the heat transfer medium. The energy balance equation for each heat exchanger can be expressed as the rate of energy coming in and is equal to that rate of energy leaving the component, which is shown in (11.13):

$$\dot{Q}_{\text{ch},i} = \dot{m}_c C_{p,i} \left(T_{\text{out},c,i} - T_{\text{in},c,i+1} \right) = \dot{m}_{h,w} C_{p,w} \left(T_{\text{out},w} - T_{\text{in},w} \right) \tag{11.13}$$

where

$\dot{Q}_{\text{ch},i}$ = rate of heat transfer during charging
$\dot{m}_c$ = mass flow rate of the working fluid
$C_{p,i}$ = specific heat of the working fluid
$\dot{m}_{h,w}$ = mass flow rate of the heat transfer medium
$C_{p,w}$ = specific heat of the heat transfer medium.

In this case, w denotes the heat transfer medium, which is also air. $T_{in,w}$ denotes the temperature of incoming cooling fluid, equivalent to the ambient temperature. The heat exchanger effectiveness is expressed as ratio of actual heat transfer to the maximum possible heat transfer [36]; (11.14) is the ratio deduced for the current work:

$$\varepsilon_i = \frac{T_{out,c,i} - T_{in,c,i+1}}{T_{out,c,i} - T_{in,w}} \tag{11.14}$$

There will also be pressure loss in the heat exchangers and other components, as well. In order to facilitate the exergy analyses under unavoidable conditions, the pressure loss ratio is considered as the ratio of pressure lost by the component to the pressure input to the component [36].

11.3.4 Storage tanks

The lowest pressure of the air storage is constrained by the operating pressure of the air motor; thus, some amount of air remains in the reservoir. The air mass that could be stored is constrained by the maximum pressure permitted in the storage tank and is calculated by applying the concept of ideal gas law. Therefore, mass stored is deduced as the ratio of product of pressure difference created in the tanks and volume of the tank to the temperature going into the tanks as shown in (11.15):

$$m_s = N_{res} \frac{(P_{max} - P_{min})V_{res}}{r \cdot T_{in,res}} \tag{11.15}$$

where
    m_s = air mass stored in reservoir
    N_{res} = number of reservoirs
    P_{max}, P_{min} = maximum and minimum pressures in the reservoir
    V_{res} = volume of the reservoir
    $T_{in,res}$ = input temperature of the reservoir.

In addition, there is also some amount of air leakage from the air storage tanks, which is largely unavoidable. Thus, the air leakage in the storage tanks can be determined as the mass of air leaked out of the accumulator with respect to those remains, as shown in (11.16) [38]. Finally, the time required to completely charge is determined as the ratio of mass stored to the rate of mass flow of the working fluid which can be expressed as (11.17).

$$\omega = \frac{m_{leak}}{m_{in,acc}} \tag{11.16}$$

$$t = \frac{m_s}{\dot{m}_c} \tag{11.17}$$

There are a few assumptions made during the storage phase, and they are as follows:

1. The temperature gradient inside the storage tank is insignificant.
2. The heat capacity of the compressed air is constant, independent of the pressure variation.
3. The thermal resistance of the wall thickness is insignificant compared to the resistance due to natural convection.

11.3.5   Turbine

The discharge time is calculated as ratio of mass stored to the mass flow rate of working fluid in expansion phase which can be seen in (11.18) [13]:

$$t_{\text{dis}} = \frac{m_s}{\dot{m}_e} \tag{11.18}$$

For small-scale applications, an open research question remains over the best expander to choose between volumetric (piston) and high-speed axial turbines [39]. Volumetric expanders like scroll machines exhibit high performance but are restricted to small pressure variations [39]. Piston expanders are advantageous for most CAES systems where high pressures would be expected. Cheayb *et al.* opted for piston expanders; hence, we also considered the same type. Through correlation with compression, the thermodynamic power and the output temperature of the ideal expansion cycle are obtained from (11.19) and (11.20):

$$T_{\text{out},e} = T_{\text{in},e} \cdot \beta_d^{\frac{1-n_e}{n_e}} \tag{11.19}$$

$$\dot{W}_{\text{th},e} = \frac{\gamma - 1}{\gamma} \cdot \frac{n_e - 1}{n_e} \cdot \dot{m}_e \cdot T_{\text{in},e} \cdot C_{p,a} \cdot \left[1 - \beta_d^{\frac{1-n_e}{n_e}} \right] \tag{11.20}$$

where β_d = expansion ratio in the turbine
$T_{\text{in},e}$, $T_{\text{out},e}$ = input and output temperatures of the compressor
$\dot{W}_{\text{th},e}$ = rate of work done by the turbine
$\dot{m}_e$ = mass flow rate through the turbine
n_e = polytropic coefficient
γ = specific heat ratio or adiabatic index or heat capacity index.

Both the input pressure and the temperature of the expansion valve are equal to the values of the storage tank and decrease with time. However, the current work is limited to steady-state conditions only. The thermodynamic or pneumatic mechanical efficiency is introduced to account for the deviation between actual and ideal thermodynamic cycles and mechanical losses. This efficiency can be expressed as the ratio of mechanical shaft power to the thermodynamic power as shown in (11.21):

$$\eta_{\text{th}} = \eta_m \cdot \eta_{\text{th}} = \frac{\dot{W}_{m,e}}{\dot{W}_{\text{th},e}} \tag{11.21}$$

where

η_{th} = thermodynamic efficiency
η_m = mechanical efficiency
$\dot{W}_{m,e}$ = mechanical shaft power
$\dot{W}_{th,e}$ = thermodynamic power.

11.3.6 Thermodynamic (exergy) analyses

To analyze the performance of the T-CAES system comprehensively, exergy analyses are executed to determine the sources, location, and extent of the exergy
destruction. The exergy of a system at a certain thermodynamic state is the maximum amount of work that can be obtained when the system moves from that
particular state to a state of equilibrium with its surroundings; exergy losses relate
to lost work. In order to perform these analyses, an exergy balance equation is built
for each component [40]. The mass balance, energy, and exergy balance equations
for each component of a system can be written as shown in (11.22)–(11.24):

$$\text{Mass balance}: \sum \dot{m}_{in} - \sum \dot{m}_{out} = 0 \tag{11.22}$$

$$\text{Energy balance}: Q - W = \sum \dot{m}_{out} \cdot h_{out} - \sum \dot{m}_{in} \cdot h_{in} \tag{11.23}$$

$$\text{Exergy balance}: Q - W = \sum \dot{m}_{out} \cdot e_{out} - \sum \dot{m}_{in} \cdot e_{in} + E_D \tag{11.24}$$

where Q is the heat transfer rate (kW) to the control volume, W is the rate of
work leaving the control volume (kW), m is the mass flow rate (kg/s), h is the
specific enthalpy (kJ/kg), e is the specific exergy (kJ/kg), and E_D is the exergy
destruction rate.

Thermomechanical exergy and chemical exergy are the two terms to express
the overall exergy of the system [3]. Since our system has no chemical reaction, the
chemical exergy is zero. The thermomechanical exergy is the extreme amount of
effective energy. Therefore, the total exergy of the fluid stream is equal to the
product mass flow rate and specific exergy which is shown in (11.25):

$$\dot{E} = \dot{m} \cdot e \tag{11.25}$$

The expression for specific exergy is as given in (11.26):

$$e = (h - h_0) - T_0(s - s_0) \tag{11.26}$$

where h_0 and s_0 are the specific enthalpy and specific entropy, respectively, at the
ambient environmental condition. T_0 is the environment temperature. We need
temperature T and pressure P under actual conditions to determine enthalpy,
entropy, and exergy. These parameters for actual conditions are taken from Cheayb
[13], in which these are determined by experimental setup.

The same parameters are calculated for the unavoidable working conditions by
considering additional assumptions, like efficiencies and losses in the system, as
shown in Table 11.8.

Table 11.8   Parametric considerations

Components	Parameters	Actual conditions [13]	Unavoidable conditions
CS1	Isentropic efficiency	85%	95% [35,37]
	Compression ratio	7.0	8.0
CS2	Isentropic efficiency	85%	95% [35,37]
	Compression ratio	7.0	8.0
CS3	Isentropic efficiency	85%	95%
	Compression ratio	7.0	8.0
HX1	Effectiveness	0.58	0.9 [35,36]
	Air pressure loss	1.2 kPa	$0.38 \cdot \Delta p_{\text{actual}}$ [6]
HX2	Effectiveness	0.796	0.9 [35,36]
	Air pressure loss	1.2 kPa	$0.38 \cdot \Delta p_{\text{actual}}$ [6]
HX3	Effectiveness	0.83	0.9 [35,36]
	Air pressure loss	1.2 kPa	$0.38 \cdot \Delta p_{\text{actual}}$ [6]
ACC	Air leak ratio	35%	10%
PR	Pressure loss ratio	0.98	0.3
	Air leak ratio	0.1	0.02
T	Isentropic efficiency	85%	95% [13]

11.3.7   Analysis methodology

The exergy equations were developed for each component based on the fuel and product concept. For the complete conventional exergy analyses of the kth component of the small-scale T-CAES system, the variables such as exergy destruction, exergy efficiency, and exergy destruction ratio are delineated in (11.27)–(11.29). Here, exergy destructed is defined as the difference between exergy of fuel and the exergy of product, exergy efficiency is defined as the ratio of exergy of product to the exergy of fuel, and destruction ratio is defined as ratio of amount of exergy destructed to the exergy of fuel.

$$E_D = E_{F,K} - E_{P,K} \tag{11.27}$$

$$E_{\text{eff}} = \frac{E_{P,K}}{E_{F,K}} \tag{11.28}$$

$$y_k = \frac{E_{D,K}}{E_{F,K}} \tag{11.29}$$

In the above equations, E_D is the amount of exergy destroyed, $E_{F,K}$ is the exergy of fuel, and $E_{P,K}$ is the exergy of the product. Also, y_k and E_{eff} are the exergy destruction ratio and exergy efficiency, respectively. To determine E_D, the energy and exergy balances are essential. In Table 11.9, these equations are delineated for each component of the small-scale T-CAES systems. The total exergy efficiency of the system is determined as the ratio of total product exergy to the total amount exergy destroyed on fuel side which is shown in (11.30):

$$\varepsilon_{\text{tot}} = \frac{E_{P,\text{tot}}}{E_{F,\text{tot}}} \tag{11.30}$$

Table 11.9 Energy and exergy balance equations for the components of the system

Components	Energy balance equations	Exergy balance equations $E_{D,\ K} = (E_{F,K}) - (E_{P,K})$
CS1	$W_{CS1} = \dot{m}_1 (h_2 - h_1)$	$E_{D,\ K} = (E_{12}) - (E_2 - E_1)$
HX1	$\eta_{HX1} \cdot \dot{m}_2 (h_2 - h_3) = \dot{m}_{15} (h_{16} - h_{15})$	$E_{D,\ K} = (E_2 - E_3) - (E_{16} - E_{15})$
CS2	$W_{CS2} = \dot{m}_3 (h_4 - h_3)$	$E_{D,\ K} = (E_{13}) - (E_4 - E_3)$
HX2	$\eta_{HX2} \cdot \dot{m}_4 (h_4 - h_5) = \dot{m}_{17} (h_{18} - h_{17})$	$E_{D,\ K} = (E_4 - E_5) - (E_{18} - E_{17})$
CS3	$W_{CS3} = \dot{m}_5 (h_6 - h_5)$	$E_{D,\ K} = (E_{14}) - (E_6 - E_5)$
HX3	$\eta_{HX3} \cdot \dot{m}_6 (h_6 - h_7) = \dot{m}_{19} (h_{20} - h_{19})$	$E_{D,\ K} = (E_6 - E_7) - (E_{20} - E_{19})$
ACC	$\dot{m}_7.h_7 = \dot{m}_8.h_8$	$E_{D,\ K} = E_7 - E_8$
PR	$\dot{m}_8.h_8 = \dot{m}_9.h_9$	$E_{D,\ K} = E_8 - E_9$
T	$W_{TUR} = \dot{m}_9 (h_{10} - h_9)$	$E_{D,\ K} = (E_{10} - E_9) - (E_{11})$

In the above equation, $E_{p,tot}$ is the exergy delivered to the grid and $E_{F,tot}$ is the total exergy entering the system. Also, a better understanding of the components is achieved by comparing the performance of the system under one condition with its performance under conditions that are altered slightly. Through this process, we can better see which components are more sensitive and effective for the efficiency improvement. Equation (11.31) for relative exergy destruction is used for comparing the results under actual and unavoidable working conditions:

$$Ex_{rel} = \frac{E_{D,K2} - E_{D,K1}}{0.5 \cdot \left(E_{D,K2} + E_{D,K1}\right)}$$ (11.31)

where Ex_{rel} = relative exergy destruction
E_{D,K_1} = exergy destruction of the component at state condition 1
E_{D,K_2} = exergy destruction of the component at state condition 2.

11.4 Results and discussions

This chapter describes the major results from the analyses and the conclusions that can be formed from them. The following section also discusses the recommendations and future works that can be considered.

The thermodynamic performance (exergy) analyses have been carried out for the proposed system. The main thermodynamic properties under actual and unavoidable working conditions are shown in Table 11.10. Also, in Table 11.11, thermodynamic properties are given for state point 2 (where the parameters differ slightly from the properties under actual and unavoidable conditions in state point 1). Two state points are considered for each condition in order to determine which component is more sensitive to the changes. Even though a specific component exhibits a higher exergy destruction, it might be another component that is showing more sensitivity to the changes in its conditions. For the system considered here, the exergy analyses under actual conditions shows that more than half of the total exergy destruction is caused by the accumulator. This loss occurs over both the charge and

Table 11.10   Parameters at first condition calculated using (11.11)–(11.20)

State point	Actual conditions			Unavoidable conditions		
	T (K)	P (bar)	m (kg/s)	T (K)	P (bar)	m (kg/s)
1	296.5	1.25	0.0039	296.5	1.25	0.0035
2	391	7.7	0.0039	386.2	8.1	0.0035
3	335	6.5	0.0039	304.12	7.6	0.0035
4	471.7	44.4	0.0039	434.9	45.6	0.0035
5	331	43.2	0.0039	310.65	45.1	0.0035
6	425	315	0.0039	399.9	320.2	0.0035
7	306	313.8	0.0039	305.4	319.4	0.0035
8	297	300	0.015	303.4	313.0	0.0189
9	183	5	0.0136	300.3	219.1	0.0185
10	281.52	1.013	0.0136	281.4	54.7	0..0185
15	295	1.013	0.0055	295	1.013	0.0044
16	334.7	1.013	0.0055	359.7	1.013	0.0044
17	295	1.013	0.0055	295	1.013	0.0044
18	394.3	1.013	0.0055	392.2	1.013	0.0044
19	295	1.013	0.0055	295	1.013	0.0044
20	379.3	1.013	0.0055	369.6	1.013	0.0044

Table 11.11   Parameters at second condition calculated using (11.11)–(11.20)

State point	Actual conditions			Unavoidable conditions		
	T (K)	P (bar)	m (kg/s)	T (K)	P (bar)	m (kg/s)
1	296.5	1.25	0.0043	296.5	1.25	0.0041
2	394	11.4	0.0043	389.2	12	0.0041
3	338.1	10.2	0.0043	307.12	11.5	0.0041
4	474.1	69.4	0.0043	436.9	69.2	0.0041
5	33.9	68.2	0.0043	313.7	68.7	0.0041
6	427.9	497.5	0.0043	402.9	488.1	0.0041
7	309	496.3	0.0043	308.4	487.5	0.0041
8	300	474.5	0.016	306.4	477.7	0.0225
9	286	9.48	0.0144	303.3	334.4	0.022
10	284.5	1.5	0.0144	284.4	83.5	0..022
15	298	1.5	0.0055	298	1.5	0.0044
16	341.7	1.5	0.0055	374.5	1.5	0.0044
17	298	1.5	0.0055	298	1.5	0.0044
18	407.5	1.5	0.0055	412.8	1.5	0.0044
19	298	1.5	0.0055	298	1.5	0.0044
20	390.9	1.5	0.0055	386.1	1.5	0.0044

discharge phases. About a quarter of the destruction is caused by the pressure regulator and turbine. Further, the analysis under unavoidable conditions reveals that the
pressure regulator, turbine, and accumulator offer 65.7%, 32.3%, and 27% recoverable exergies, respectively; these contribute to a 35% increase in overall exergy

efficiency of the system. This result indicates that there is great potential for improvement. The results for the exergy analyses of the proposed T-CAES system under actual and unavoidable working conditions are delineated in Tables 11.12 and 11.13. These results were computed in MATLAB® using (11.27)–(11.30).

The tabled results confirm the dominant role played by the accumulator in the sum of exergy destruction (more than half). This loss is due to significant air leakage in the high-pressure, low-temperature storage tanks. Nearly 35% of air leakage is happening under actual conditions. Under unavoidable working conditions, the storage tanks also exhibit high exergy destruction. The fuel and product exergies associated with the accumulator are relatively higher than the other components in the system, which can be seen in Table 11.12. The second and third highest exergy destruction rates occur in the pressure regulator and the turbine, respectively, with corresponding values of 2.39 and 1.3 kW. The high exergy destruction rate in the turbine is due to the low isentropic efficiency in the low-pressure turbine; significant losses from the pressure regulator also notably contribute to exergy destruction. Meanwhile, under unavoidable conditions, the exergy

Table 11.12   *Results of the exergy analyses of small-scale T-CAES under actual working conditions calculated using (11.27)–(11.30)*

Component	$\dot{E}_{F,\,K}$ (kW)	$\dot{E}_{P,\,K}$ (kW)	$\dot{E}_D$ (kW)	$\dot{E}_{D\_rel}$ (kW)	E_{eff} (%)	Y (%)
CS1	1.01	0.722	0.287	0.11	70	28.4
CS2	1.06	0.778	0.238	0.31	72.9	22.4
CS3	1.1	0.737	0.363	0.11	66.9	33
HEX1	0.041	0.013	0.027	0.88	33	65.8
HEX2	0.145	0.077	0.067	0.07	53	46.2
HEX3	0.088	0.057	0.031	0.06	64	35.2
ACC	81	78.2	2.87	0.02	21	35.4
PR	7.3	4.91	2.39	0.03	26.6	73.9
TUR	1.86	0.538	1.3	0.51	28	69.8

Table 11.13   *Results of the exergy analyses of small-scale T-CAES under unavoidable working conditions calculated using (11.27)–(11.30)*

Component	$\dot{E}_{F,\,K}$ (kW)	$\dot{E}_P$ (kW)	$\dot{E}_D$ (kW)	$\dot{E}_{D\_rel}$ (kW)	E_{eff} (%)	Y (%)
CS1	0.936	0.659	0.277	0.16	71	29.6
CS2	0.829	0.620	0.195	0.18	78	21.7
CS3	0.931	0.634	0.293	0.18	69.6	30.3
HEX1	0.058	0.028	0.03	0.28	46	53.3
HEX2	0.107	0.059	0.04	0.24	64.7	34.7
HEX3	0.062	0.036	0.015	0.32	66.9	33.3
ACC	81.23	79.6	1.63	0.08	42.3	10.3
PR	9.31	8.55	0.763	0.19	51	8.2
TUR	2.19	1.37	0.82	0.18	53.4	37.6

Exergy efficiencies

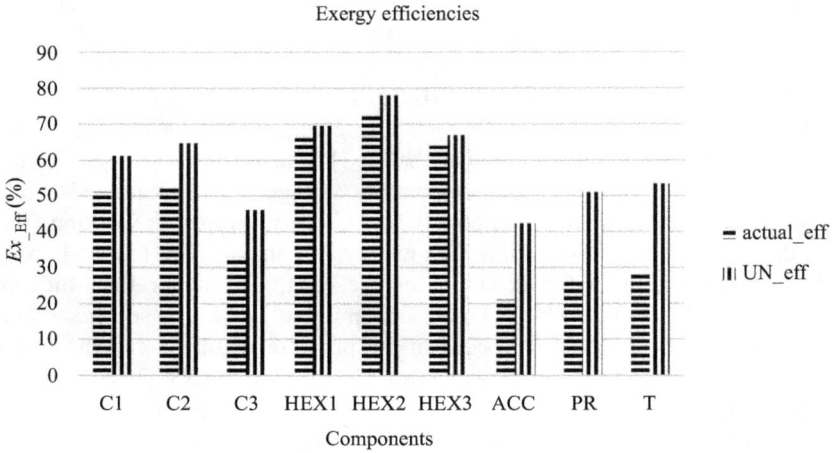

Figure 11.11   Figure illustrating the exergy efficiencies under two working conditions

Table 11.14   Results of accumulator

Component	Exergy destructed		Exergy efficiency	
	$\dot{E}_D$ (kW)	$\dot{E}_{D\_UN}$ (kW)	E_{eff} (%)	E_{eff_un}(%)
Acc	2.87	1.63	21	42.3

destruction rates of these components decreased to levels below the actual condition values. There is not much difference in the component-wise order of destruction rates, as it follows the same order as in actual working conditions.

In terms of exergy efficiencies—shown in Figure 11.11—though the accumulator shows the highest amount of exergy destruction, it is the pressure regulator that demonstrates a great potential for improvement. This is revealed by the exergy efficiency of the pressure regulator increasing from 26.6% under the actual condition to 51% under the unavoidable condition. It is clear that the exergy efficiencies of the components under unavoidable conditions are higher than those under actual conditions. The compressors and the pressure regulator exhibit higher exergy efficiencies than the heat exchangers, even though they have high exergy destruction rates when compared to heat exchangers. This finding indicates that the amount of fuel exergy destroyed in the heat exchangers is relatively higher than the exergy destroyed on the product side. The increase in overall system exergy efficiency from 17% to 51.3% indicates significant potential for improvement.

The results in Table 11.14 confirm the dominant role of the accumulator in the sum of exergy destruction, as it was responsible for more than half. This loss is due to significant air leakage in the high-pressure, low-temperature storage tanks. Nearly 35% of air leakage is happening under actual conditions. Under unavoidable working conditions, the storage tanks also exhibit high exergy destruction. The fuel and product exergies associated with the accumulator are relatively higher than the other components in the system.

The relative exergy destruction determined using (11.31) for different components of the small-scale T-CAES system under actual and unavoidable working conditions are shown in Figure 11.12. Relative exergy destruction helps us to identify which component is sensitive among multistage components and also in the overall system to the changes imposed. In the present work, we have multistage compressors and the heat exchangers. Among the three compressors, Stage 2 compressor displays more sensitivity followed by Stage 3 and 1 compressors under actual working conditions, whereas under unavoidable conditions, Stage 2 and 3 compressors show similar sensitivity. Among the heat exchangers, the order of sensitivity under actual conditions is HEX1, HEX2, and HEX3 with the destruction values of 0.88, 0.07, and 0.04. But the order changes under unavoidable conditions with HEX3 taking the first position followed by HEX1 and HEX2, respectively. When we consider the entire system, turbine is also showing higher sensitivity to the changes given.

In terms of exergy destruction ratios under actual working conditions—as shown in Figure 11.13—the pressure regulator has the highest exergy destruction ratio (73.6%) under actual conditions. This means that 73.6% of the exergy of fuel entering the pressure regulator is destroyed through direct contact with the environment. The turbine and heat exchangers 1 and 2 have the second, third, and fourth highest destruction ratios under actual working conditions, with values of 69%,

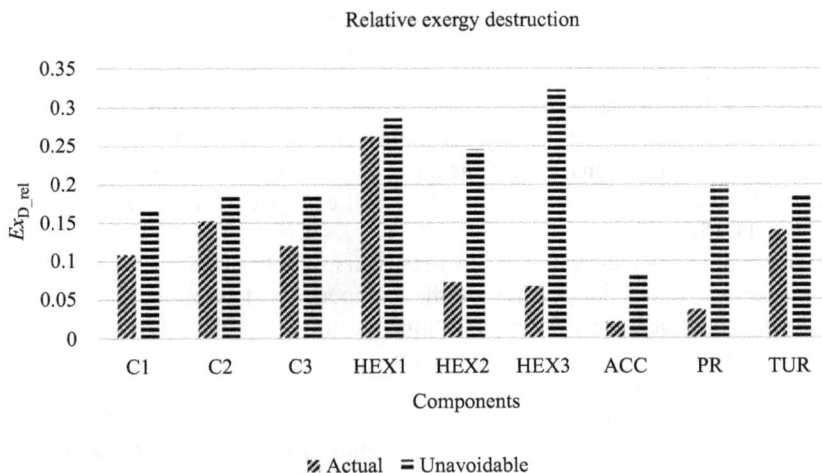

Relative exergy destruction

Figure 11.12   Figure illustrating the relative exergy under two working conditions

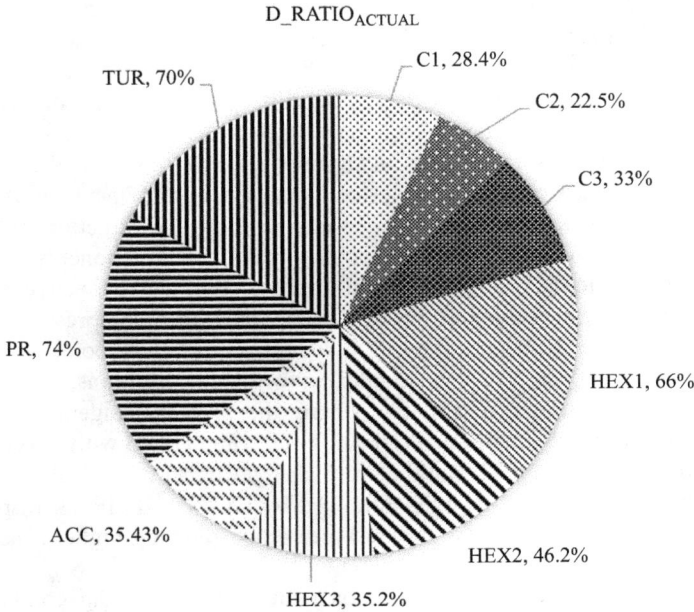

Figure 11.13   Exergy destruction ratio under actual conditions

65.8%, and 46%, respectively. High exergy destruction ratios in heat exchangers are due to large temperature differences between hot and cold streams. Under unavoidable conditions, there is a change in the order of the destruction ratios illustrated in Figure 11.14. Here, heat exchanger 1 has the highest exergy destruction ratio with 63.5%, followed by the turbine with 38% and heat exchanger 2 with 37.3%. The pressure regulator's destruction ratio has reduced from 73.6% to 8.9%, showing notable opportunity for improvement. This result tells us that significant exergy can be conserved if the air leakage and pressure losses are reduced. There is also a significant decrease in the overall exergy destruction ratio, as shown in Figure 11.15.

Finally, improvement priority order is given in Table 11.15. This order is based on the exergy destruction values of the components determined using defined equations. Under actual working conditions, accumulator should be given the top priority, followed by the pressure regulator and the turbine; whereas, under unavoidable working conditions, the first priority is taken by accumulator again followed by a slight change as turbine displayed higher destruction than pressure regulator. Hence, the second priority is to turbine followed by pressure regulator and the compressors' stages 3, 1, and 2. From this result, we can improve our actual focusing more on the top priority components.

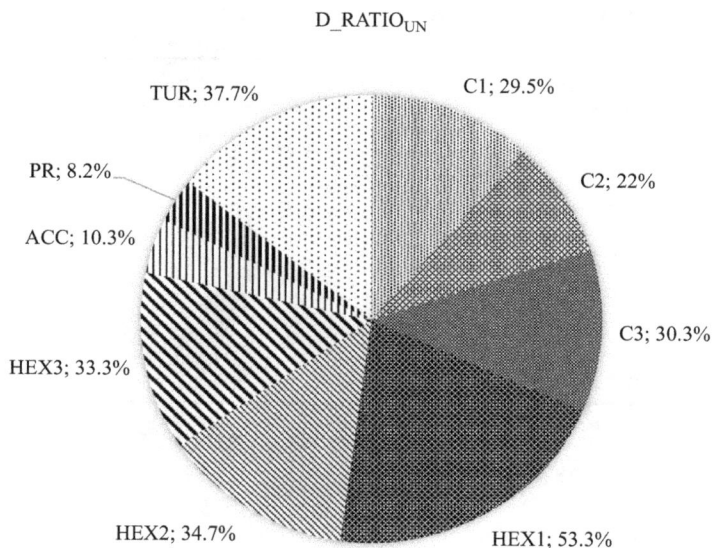

Figure 11.14   Exergy destruction ratio under unavoidable conditions

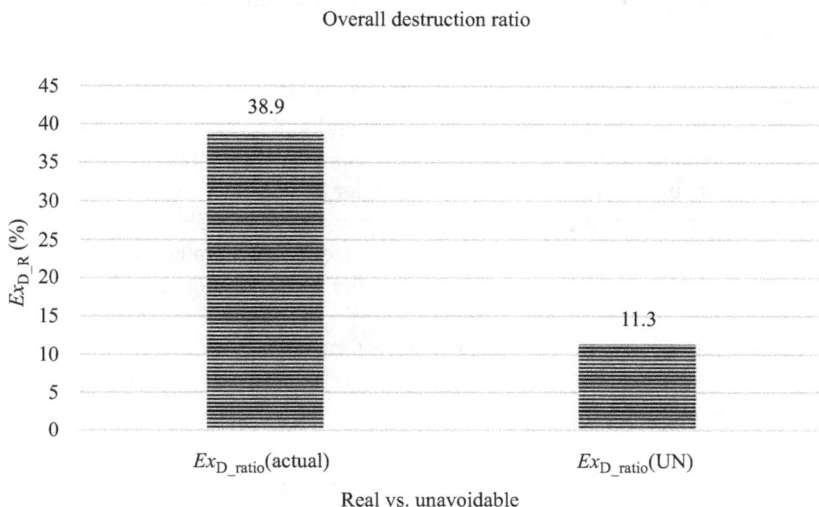

Figure 11.15   Comparison of overall exergy destruction ratio of the system under two working conditions

Table 11.15   Improvement priority order for the components

Rank	Actual conditions		Unavoidable conditions	
	Component	E_D	Component	E_D
1	HEX1	0.287	HEX3	0.277
2	TUR	0.238	HEX1	0.195
3	CS2	0.363	HEX2	0.293
4	CS3	0.027	PR	0.03
5	CS1	0.067	CS3	0.04
6	HEX2	0.031	TUR	0.015
7	HEX3	2.87	CS2	1.63
8	PR	2.39	CS1	0.763
9	ACC	1.3	ACC	0.82

11.5   Conclusions

Exergy analyses are performed on a small-scale (4 kW compressor and 2 kW turbine) T-CAES system, which was proposed in experimental form by Cheayb *et al.* [13]. Our principal conclusions are as follows.

Results suggest that, in the proposed T-CAES system, the individual components offer a significant number of recoverable exergies, which have increased the overall exergy efficiency of the system under consideration by 35%.

- The largest quantity of exergy is destroyed in the accumulators under both actual and unavoidable working conditions. There is a significant reduction, however, in the amount of exergy destroyed in accumulators from the actual to the unavoidable conditions.
- Even though the accumulators show high exergy destruction, it was revealed that, in terms of destruction ratio, the pressure regulator shows the highest (73.1%) under actual working conditions and 8.2% under unavoidable conditions. These numbers suggest that the pressure regulator has a high fuel exergy destruction under actual conditions. Under unavoidable conditions, though, heat exchanger 1 displays a high destruction ratio of 63.5%. With this finding, we can conclude that, even though a specific component exhibits a high destruction rate, it will not necessarily be the same component with the highest potential for improvement.
- Calculating relative exergy destruction of each of the *k*th components of the system-enabled systematic sensitivity analyses that helped determine which components would be most affected by parametric change.

Future work will look to optimize the system configuration and actual parameters. A steady-state condition was considered for this system, as it is informative and straightforward to implement, but can be limiting when the compression ratio varies. In our future work, the transient nature of the discharge phase will be accounted for to better assess the cooling potential and the output power. We will also execute more detailed, "advanced" exergy analyses. This will improve the identification of changes required to improve the system performance. Avoidable and unavoidable conditions will be considered in these advanced analyses. Finally, a few potential improvement strategies, mentioned in the following, could be used in future work to reduce the losses in the pressure regulator, which, in turn, improves the overall system performance.

- Increasing the expansion ratio to 25 by introducing the recently developed microturbines [36].
- Replacing the throttling valve with a Ranque–Hilsch vortex tube or a cascade of vortex tubes. These devices have the capability to relax the air to reduce the maximum inlet pressure allowed by current microturbines, while producing a hot and a cold stream. The produced cooling power would be then reinjected within some heat exchangers of the CAES system toward improved efficiency.

References

[1] Luo, X., Wang, J., Dooner, M., and Clarke, J. (2015). Overview of current development in electrical energy storage technologies and the application potential in power system operation. *Applied Energy*, 137, 511–536.

[2] Global Energy Statistical Yearbook 2020. (2020). Enerdata. https://yearbook.enerdata. net/world4electricity-production-map-graph-and-data.html.

[3] Barbour, E., Mignard, D., Ding, Y., and Li, Y. (2015). Adiabatic compressed air energy storage with packed bed thermal energy storage. *Applied Energy*, 155, 804–815.

[4] Budt, M., Wolf, D., Span, R., and Yan, J. (2016). A review on compressed air energy storage: Basic principles, past milestones, and recent developments. *Applied Energy*, 170, 250–268.

[5] Chen, H., Cong, T. N., Yang W., Tan, C., Li, Y., and Ding, Y. (2009). Progress in electrical energy storage system: A critical review. *Progress in Natural Science*, 19(3), 291–312.

[6] Ebrahimi, M., Carriveau, R., Ting, D. S., and McGillis, A. (2019). Conventional and advanced exergy analyses of a grid connected underwater compressed air energy storage facility. *Applied Energy*, 242, 1198–1208.

[7] Guo, C., Pan, L., Zhang, K., Oldenburg, C. M., Li, C., and Li, Y. (2016). Comparison of compressed air energy storage process in aquifers and caverns based on the Huntorf CAES plant. *Applied Energy*, 181, 342–356.

[8] Luo, X., Wang, J., Krupke, C., et al. (2016). Modelling study, efficiency analyses and optimization of large-scale adiabatic compressed air energy storage systems with low-temperature thermal storage. *Applied Energy*, 162, 589–600.

[9] Yu, Q., Wang, Q., Tan, X., Fang, G., and Meng, J. (2019). A review of compressed-air energy storage. *Journal of Renewable and Sustainable Energy*, 11(4), 042702.

[10] Bullough, C., Gatzen, C., Jakiel, C., Koller, M., Nowi, A., and Zunft, S. (2004). Advanced adiabatic compressed air energy storage for the integration of wind energy. In *Proceedings of the European Wind Energy Conference (EWEC)* Vol. 22, p. 25.

[11] Li, Y., Wang, X., Li, D., and Ding, Y. (2012). A trigeneration system based on compressed air and thermal energy storage. *Applied Energy*, 99, 316–323.

[12] Kim, Y. M. and Favrat, D. (2010) Energy and exergy analyses of a micro-compressed air energy storage and air cycle heating and cooling system. *Energy*, 35(1), 213–220.

[13] Cheayb, M., Gallego, M. M, Tazerout, M., and Poncet, S. (2019). Modelling and experimental validation of a small-scale trigenerative compressed air energy storage system. *Applied Energy*, 239, 1371–1384.

[14] Weiqing, X., Ziyue, D., Xiaoshuang, W., Maolin, C., Guanwei, J., and Yan, S. (2020). Isothermal piston gas compression for compressed air energy storage. *International Journal of Heat and Mass Transfer*, 155, 119779.

[15] Liu, W, Liu, L, Zhou, L, et al. (2014). Analysis and optimization of a compressed air energy storage-combined cycle system. *Entropy*, 16(6), 3103–3120.

[16] Zhao, P., Wang, J., and Dai, Y. (2015). Thermodynamic analysis of an integrated energy system based on compressed air energy storage (CAES) system and Kalina cycle. *Energy Conversion and Management*, 98, 161–172.

[17] Wolf, D. and Budt, M. (2014). LTA-CAES—A low-temperature approach to adiabatic compressed air energy storage. *Applied Energy*, 125, 158–164.

[18] Zhang, Y., Yang, K., Li, X., and Xu, J. (2013). The thermodynamic effect of thermal energy storage on compressed air energy storage system. *Renewable Energy*, 50, 227–235.

[19] Xia, C., Zhou, Y., Zhou, S., Zhang, P., and Wang, F. (2015). A simplified and unified analytical solution for temperature and pressure variations in compressed air energy storage caverns. *Renewable Energy*, 74, 718–726.

[20] Mozayeni, H., Negnevitsky, M., Wang, X., Cao, F., and Peng, X. (2017). Performance study of an advanced adiabatic compressed air energy storage system. *Energy Procedia*, 110, 71–76.

[21] He, Y., Chen, H., Xu, Y., and Deng, J. (2018). Compression performance optimization considering variable charge pressure in an adiabatic compressed air energy storage system. *Energy*, 165, 349–359.

[22] Houssainy, S., Janbozorgi, M., and Kavehpour, P. (2018). Thermodynamic performance and cost optimization of a novel hybrid thermal-compressed air energy storage system actual. *Journal of Energy Storage*, 18, 206–217.

[23] Kim, Y. M. (2012). Novel concepts of compressed air energy storage and thermo-electric energy storage. EPFL.

[24] Mazloum, Y., Sayah, H., and Nemer, M. (2017). Dynamic modeling, and simulation of an isobaric adiabatic compressed air energy storage (IA-CAES) system. *Journal of Energy Storage*, 11, 178–190.

[25] Bagdanavicius, A. and Jenkins, N. (2014). Exergy and exergo-economic analyses of a compressed air energy storage combined with a district energy system. *Energy Conversion and Management*, 77, 432–440.

[26] Geissbühler, L., Becattini, V., Zanganeh, G., et al. (2018). Pilot-scale demonstration of advanced adiabatic compressed air energy storage, Part 1: Plant description and tests with sensible thermal-energy storage. *Journal of Energy Storage*, 17, 129–139.

[27] Becattini, V., Geissbühler, L., Zanganeh, G., Haselbacher, A., and Steinfeld, A. (2018). Pilot-scale demonstration of advanced adiabatic compressed air energy storage, Part 2: Tests with combined sensible/latent thermal-energy storage. *Journal of Energy Storage*, 17, 140–152.

[28] Carriveau, R., Ebrahimi, M., Ting, D. S., and McGillis, A. (2019). Transient thermodynamic modeling of an underwater compressed air energy storage plant: Conventional versus advanced exergy analyses. *Sustainable Energy Technologies and Assessments*, 31, 146–154.

[29] Arabkoohsar, A., Dremark-Larsen, M., Lorentzen, R., and Andresen, G. B. (2017). Subcooled compressed air energy storage system for coproduction of heat, cooling, and electricity. *Applied Energy*, 205, 602–614.

[30] Lv, S., He, W., Zhang, A., Li, G., Luo, B., and Liu, X. (2017). Modelling and analyses of a novel compressed air energy storage system for trigeneration based on electrical energy peak load shifting. *Energy Conversion and Management*, 135, 394–401.

[31] Liu, J. L. and Wang, J. H. (2015). Thermodynamic analyses of a novel tri-generation system based on compressed air energy storage and pneumatic motor. *Energy*, 91, 420–429.

[32] Han, Z. and Guo, S. (2018). Investigation of discharge characteristics of a tri-generative system based on advanced adiabatic compressed air energy storage. *Energy Conversion and Management*, 176, 110–122.

[33] Jannelli, E., Minutillo, M., Lavadera, A. L., and Falcucci, G. (2014). A small-scale CAES (compressed air energy storage) system for stand-alone renewable energy power plant for a radio base station: A sizing-actual methodology. *Energy*, 78, 313–322.

[34] Venkataramani, G., Ramakrishnan, E., Sharma, M. R., et al. (2018). Experimental investigation on small capacity compressed air energy storage towards efficient utilization of renewable sources. *Journal of Energy Storage*, 20, 364–370.

[35] Tsatsaronis, G. and Park, M. H. (2002). On avoidable and unavoidable exergy destructions and investment costs in thermal systems. *Energy Conversion and Management*, 43(9-12):1259–1270.

[36] Shah, R. K. and Sekulic, D. P. (2003). *Fundamentals of Heat Exchanger Design*. Hoboken, NJ: John Wiley & Sons.

[37] Bloch, H. P. (2006). *A Practical Guide to Compressor Technology*. John Wiley & Sons.

[38] Mohammadi, A. and Mehrpooya, M. (2017). Energy and exergy analyses of a combined desalination and CCHP system driven by geothermal energy. *Applied Thermal Engineering*, 116, 685–694.

[39] Weib, A. P. Volumetric expander versus turbine—Which is the better choice for small ORC plants. (2015). In *3rd International Seminar on ORC Power Systems*, 12–14.

[40] Wang, Z., Xiong, W., Ting, D. S., Carriveau, R., and Wang, Z. (2016). Conventional and advanced exergy analyses of an underwater compressed air energy storage system. *Applied Energy*, 180, 810–822.

Chapter 12

Offshore systems

Daniel Chidiebere Onwuchekwa[1]

Compressed air energy storage (CAES) systems can be designed such that the air is stored underwater and at high pressures in lightweight reinforced balloons called energy bags [1,2]. This chapter shows an offshore device, Buoyancy Engine, that effectively harnesses the resultant buoyant force acting on an inflated energy bag by converting the upward vertical thrust of the bag (when released) to torque with the help of a rack-and-pinion gear system. Similar to lifting bags for salvaging objects from the sea, the energy bag works by rising a certain distance through the water column during the generation stage. The system was designed such that a rack gear is attached to the energy bag at its base. The rack gear is also in mesh with a pinion gear at a given point along its length. An energy bag filled with the stored compressed air is allowed to float during the buoyancy engine operation, pulling along the rack gear, thereby causing torque to be generated in the pinion gear because the rack and pinion gears are in mesh. Furthermore, the pinion gear's torque is transferred through a shaft to an electric generator designed to be a quick-start, variable speed and protected in a waterproof housing anchored at the bottom of the sea.

The design achieves a positive network output by considering the ascent speed and the resultant forces acting on the submerged energy bag linked to a rack gear in mesh with a pinion gear. This design's calculations showed that the pinion gear produced high torque at a low angular speed for sufficient electrical power to be produced in the generator. By appropriately tuning the short-term electrical power towards higher voltages, the electricity can be used in a purpose-built electric arc furnace [3] to instantly heat a eutectic mixture of molten salts to high temperatures of over 550 °C [3,4]. For an adiabatic CAES system, some of the heat energy stored in the molten salt can be utilised along with the harnessed heat of compression in multiple generation stages to heat up the stored air for higher work output. Furthermore, similar to a concentrated solar thermal plant, the molten salt can be used as a heat transfer fluid to slowly generate electricity in a coupled heat engine such as a steam turbine [4].

[1]Independent Research Engineer

Keywords: Compressed air energy storage; offshore; underwater; buoyancy engine; thermal storage

This chapter looks to show how a different generation stage was incorporated into the integrated compressed air renewable energy systems (ICARES) [2]. Previous work on adiabatic CAES [4] showed that using only the stored heat energy captured during the compression of the air is insufficient to generate a high work output at the turbine due to heat loss. Hence, a supplementary source of heat is required. This chapter presents a renewable energy concept for the supply of the required heat energy.

The additional generation stage to the ICARES involves harnessing energy from the buoyant forces acting on the energy bag when fully inflated and submerged in water [2]. Similar to lifting bags, the energy bags will be allowed to rise a certain height along the water column.

This chapter will also look at the corresponding torque and speed on the pinion and, if required, the required gearbox ratio to efficiently generate power. The system was designed such that rack gear is fixed at one end to the base of the energy bag. As the rack is lifted along the water column by the energy bag, it meshes with a set of pinions, causing torque and rotation, which is transferred via a shaft to an electric generator.

Before the generation stage, the bulk of the rack gear fits into decommissioned and abandoned offshore oil and gas production casing bore [5] (this acts as a sump for the gear). Currently, the proposed siting of this concept is decommissioned and abandoned offshore oil and gas fields [5].

Since the energy bag moves along the column of water, the design and filling of the bag are carried out following the IMCA D 016 regulation [6] for closed underwater airlift bags. However, the buoyancy engine system was designed to operate without losses of the stored air via venting. Incorporating a buoyancy engine into an underwater CAES system means that the energy bag will be initially anchored at water depths greater than the depth required to release the air from the bags.

This chapter presents the design and components of a buoyancy engine, and a numerical model for its optimisation to determine the output power.

The following are beyond the scope of this chapter and hence will not give details of them:

1. Design of the fabric and shape of the energy bags.
2. Design and details of the required generator.
3. Details of the heat-generating system (electric arc furnace) to be used alongside the buoyancy engine.

12.1   Energy storage systems

12.1.1   Underwater energy storage

Another way CAES is being applied is the use of flexible bags to store the compressed air underwater. This system is known as the underwater CAES (UW-CAES).

The system is designed to compress and store air at a high-pressure underwater in lightweight and robust flexible bags. When the energy is needed, the pressurised air is released through pipes to be heated and expanded in a turbine to generate electricity.

Early CAES plants utilised fixed volumes like underground caverns, where the containment pressure depends on the mass of air in the containment. However, the pressure within an energy bag when submerged in deep water is isobaric [2] and independent of the mass of air present due to the constant hydrostatic pressure of the surrounding water column above the distensible bag, which could lead to energy densities as high as 25–50 MJ/m^3 [1].

The advantages of the isobaric characteristics exhibited by air in an energy bag when submerged in water cannot be overemphasised as operational CAES systems (fixed volume) decrease in efficiency as the energy density reduces. The water depth at which the bag is anchored is crucial in determining how much energy the bag can store since the depth sets the air pressure [1]. Therefore, a given energy bag achieves relatively high or low energy storage capacity in deep or shallow water, respectively. At a pressure of 100 bars, the air is far less dense than seawater of the same temperature; hence, the variation in pressure with height inside the bag is neglected.

12.1.2 Underwater inflatable lifting bags

Underwater inflatable lifting bags are air-filled fabric structures used to provide temporary support or lift objects (such as subsea installations, a sunken ship and concrete pipes) underwater, off the seabed or along the water column. Underwater inflatable lifting bags are similar to energy bags because they are both reinforced and designed to overcome similar loads. Marine contractors prefer underwater inflatable lifting bags because they are cheap, easy to deploy and reusable. The International Marine Contractors Association (IMCA) is an international trade association representing offshore and underwater engineering companies, and it does this by providing and promoting technical standards and relevant guidelines mainly through the publication of codes of practice [6]. The IMCA D 016 is currently the only international standard for underwater airlift bags. The IMCA D 016 provides information on design considerations, operational considerations, maintenance, initial and periodic examination, and certification criteria. A standard underwater airlift bag is designed with a minimum factor of safety of 5:1 on its safe working load (SWL) [6].

12.2 Buoyancy engine

The buoyancy engine is a device that converts the effective buoyant force acting on an inflated underwater energy bag to very high-voltage (dense), short-term electrical power [4]. Similar to underwater inflatable lifting bags, the energy bags are designed to ascend a certain distance through the water column during the generation stage.

This concept mechanically converts the vertical motion of the inflated underwater bag to rotational motion through sets of efficient mechanical gear systems.

A long-segmented rack gear attached on one of its ends to the base of the energy bag or buoyancy module, and in mesh with sets of pinion gears, will effectively convert the linear motion of the buoyancy module and the attached rack gear to rotational motion at the pinion gear. The low angular speed and high torque at the pinion were used to power a quick-start and variable speed electric generator. Like a bulb turbine, the components are sealed within a watertight steel housing and anchored at the bottom of the waterbody. See Figure 12.1 for an artist's impression of this concept.

The short-term and high-voltage electrical energy generated in the buoyancy engine generation stage was utilised to generate an electric arc. The arc generated is then used to heat up charged molten salt to over 500 °C [3,4] in an auxiliary heat store of the adiabatic CAES system. The high-voltage electrical power and communications are sent to the arc electrodes on a floating platform through a subsea umbilical cable.

The buoyancy engine is a concept that could be installed such that it utilises some relevant decommissioned offshore oil and gas equipment, such as production casings and suction anchors [5,7]. The system uses the decommissioned and abandoned equipment in their original installed locations, although for anchors, they may be relocated to soothe the optimum system geometry. The suction caissons primarily serve as sumps for the rack gear when the system is in a charged state. Only efficient downhole well plugging was employed for the utilised caissons.

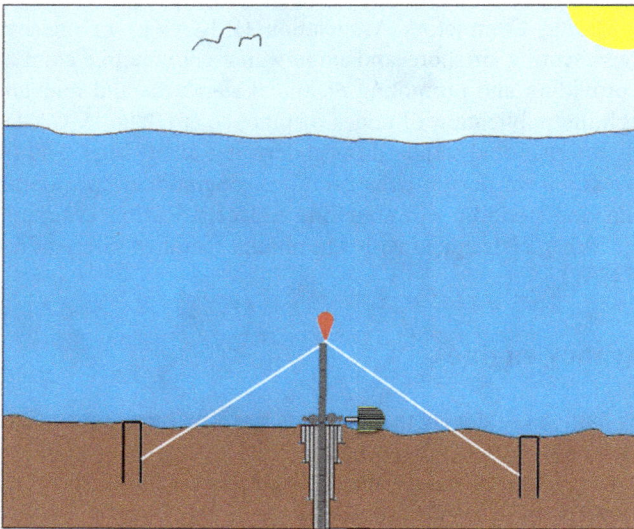

Figure 12.1   *An artist's impression of the buoyancy engine system. The heat store and power cable are not shown (image not to scale)*

The buoyancy engine consists of several distinct components, all from existing and proven technologies. They include the following:

1. Energy bag or buoyancy module:
    (a) The sealed buoyancy module containing compressed air stored for later use in power generation in an adiabatic CAES system. The heat of compression of the stored air is removed and stored separately in an efficient heat store.
    (b) The volume of the air to be stored in the buoyancy module was optimised based on
        (i)   the water depth at which the air will be released and
        (ii) the required volume to allow for sufficient buoyant force for the breaking torque of the pinion.

    (c) Large buoyancy modules (ample volume storage) and low ascent speed have been investigated and shown to allow for very high tangential forces at the pinion. The buoyancy module has an approximate vertical travel distance of 100 m.

2. Rack and pinion gears:
    (a) Rack and pinion gears made from high-grade alloy steel of which carburising, quenching and grinding were used to treat and form them. The pinion gears are built to withstand high tangential forces and torque loads and are designed to be wide for better tooth contact. The excessive weight in the pinion was removed for decreased inertia (details of gear design not given in this chapter).
    (b) The rack is segmented at regular intervals (like chains) to allow for flexibility and compensate for its high bending stiffness.
    (c) The rack is meshed on two sides (near seabed) by several pinions for maximum power output.

3. Gearbox and generator – enclosed:
    (a) Gearbox with a high gear ratio is incorporated to the speed from the pinion output shaft.
    (b) Quick-start and variable speed electric generator produce high-voltage electricity within a short period (duration of generation).

4. Power cable and control umbilical:
    (a) The power cable is used to send the high-voltage power out of the electric generator.
    (b) The electric power cable is incorporated into a subsea umbilical. The umbilical also consists of hydraulic hoses for surface control and communication cables for monitoring the subsea equipment.

5. Decommissioned and abandoned intermediate or production casings:
    (a) The preferred wellbores are those that were effectively plugged downhole. They were abandoned with the intermediate (or production casings) and without installing a well cap.

(b)   150 m deep from the seabed down the casing bore is used as a sump for the rack gear in the storage phase.

6. Suction anchor and mooring lines:
    (a)   Suction anchors used in the oil field to hold the floating structure in place provide anchor support for the buoyancy module.
    (b)   Moring lines are polyester-type mooring lines that hold the structure in place at the end of the buoyancy engine generation stage. The lightweight and neutral buoyancy of the mooring lines means that they do not exert any load on the buoyancy module except for the torsional load, which they exert at the end of the buoyancy engine generation stage.

7. Heat store
    (a)   The heat store can store heat up to 600 °C in the form of molten salt.
    (b)   Before storing the air, the heat of compression is harnessed from the air and stored either separately or in the same store as the cold molten salt.
    (c)   The short-duration high-voltage electrical energy produced in the electrical generator is efficiently used to power auxiliary heat store using an electric arc.

The buoyancy engine system can easily be scaled in decommissioned oil fields with multiple wells. The volume or quantity of the bag can easily be increased for higher tangential forces on the pinion in fields with massive suction anchors.

12.2.1   *Pressure imbalance and density variations*

The air pressure within a fully inflated bag is uniform at every point within the bag. However, this is not the case for large bags submerged underwater, as the value of the air pressure differs from the top to the bottom of the bag, and this difference increases proportionately with bag size because of the varying external hydrostatic forces. As shown in [4], this pressure imbalance accounts for minimal values between extremes of the bag and is hence negligible. Since air density is far lower than the density of water, even at extreme pressures of the air, the differences in pressure due to imbalance can be ignored. Figure 12.2 shows a trend of internal

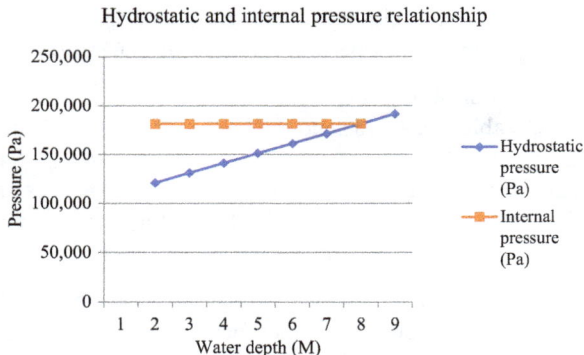

Figure 12.2   Relationship between internal pressure and hydrostatic pressure

pressures due to pressure imbalance within a bag (zero differential pressure at the base) of vertical height 6 m, anchored at 8 m in a water depth of 9 m.

For a given quantity of air at ambient temperature, the position of an inflated energy bag relative to its water depth is a crucial determinant of the buoyant forces acting on the bag. The net buoyant force per unit volume of the contained air $[(\rho_w - \rho_a)g]$ is directly proportional to the depth at which the bag is anchored. The net buoyant force reduces as the water depth increases because of the decreasing volume of the air due to the much higher compressibility of air than water.

An inflated bag not anchored to the seabed is subject to significant vertical motions due to buoyancy and will experience hydrostatic pressures acting around the bag, which vary linearly with water depth. Similarly, the air volume within the bag is continuously being altered as the density change occurs, due to the high compressibility of air. The effect of density change will be discussed further later in this chapter.

An inflated underwater bag moving vertically downward across a water column, similar to lift bags, will experience a relative volume decrease with increasing depth. The relative decrease in volume of the bag results from higher compressibility of air than water and the respective decrease in volume of the air. This occurs due to pressure balance. This decrease in the bag's volume with increasing depth results in the reduced loading of the bag, which comes from the buoyancy or upthrust generated by the volume of water displaced. Therefore, for a given volume of air, V, in an inflated underwater bag, the buoyant force acting on the bag reduces with increasing water depth.

For a bag experiencing an upward motion across a column of water, similar to closed-bottom lifting bags, it will continue to experience a relative volume increase. Hence a scaled linear increase in the buoyant force (for calm water surface conditions) is experienced. As the bag rises into lower pressures, the air within it expands. For lifting bags, the IMCA D 016 regulations [6] stipulate that for an enclosed lift bag, a relief valve releases air when the internal pressure is approximately 13.8 kPa (0.138 bar) over ambient pressure. This means that the relief valve maintains the differential pressure at approximately 13.8 kPa. A relief valve in lifting bags is a safety procedure employed to avoid overpressure within the bag, which could also cause the SWL of the bag to be exceeded on the membrane.

For compressed air storage conditions, any form of air loss from the bag will be regarded as a loss of valuable energy. When stored air is used as the buoyancy module for a buoyancy engine, the optimum volume and pressure of air required for expansion at the threshold generation (expansion in turbines) water depth (for example 500 m) must be considered as the bag will be anchored at a depth greater than the threshold for expansion generation.

Care was taken in selecting the optimum water depth and volume of the air within the bag at the anchor point before the buoyancy engine generation. This ensures that the bag displaces a sufficient volume of water to attain the required buoyancy to start the engine. Since air is not expected to be released from the energy bag before generation, the selection of the bag at the generation water depth was a crucial parameter in the design.

IMCA D 016 clearly warns against the use of fully enclosed bags where any form of ascent is planned in a statement:

> Generally, only open bottom parachute bags should be used where any form of ascent is planned or possible, such as vessel salvage or raising objects from the seabed. Fully enclosed bags should not be used for this purpose.

Nevertheless, it should be noted that major lifting operations involving closed bags have been used successfully by major manufacturers and operators of lift bags [8].

12.2.2 Buoyant forces acting on the bag

As stated above, for a unit mass of air, the effective buoyant forces acting on the bag differ for different water depth. The effective buoyancy/upthrust of an inflated underwater bag at a constant temperature for a given volume of air, V, is calculated by

$$b_g = (\rho_w - \rho_a)gV \tag{12.1}$$

The change in the buoyant force acting on the bag due to the change in volume as a result of increased hydrostatic pressure was calculated using Boyle's law. Figures 12.3 and 12.4 show how the change in density, volume and the effective buoyancy acting on the energy bag change with water depth.

12.2.3 Rack-and-pinion gear system

'*A rack and pinion system is a toothed member which moves in a straight-line path and may therefore be regarded as a portion of a wheel of infinite radius or zero curvature*' [9]. From the definition, it can be assumed that the rack is driven by the pinion and moves at a right angle to the direction of the pinion. This definition is similar to the definitions found in most modern literature [10–12], where a smaller

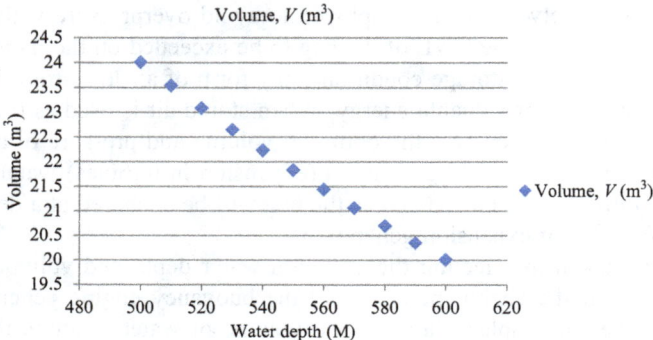

Figure 12.3 *Effect of water depth on the volume of air within the energy bag (or volume of water displaced by the bag)*

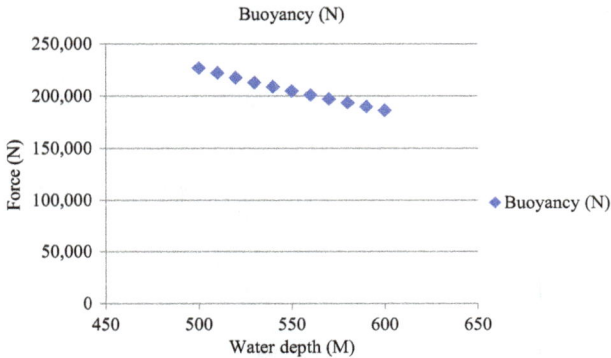

Figure 12.4   Effect of water depth on the effective buoyant force on the bag

radius pinion is always designed to drive the rack of infinite radius or zero curvature. From this, the rack-and-pinion gear system is seen to be similar to spur gears, where the rack represents the bull gear.

A pinion with an infinite radius or zero curvature driven by a rack with a known radius (a reverse of the conventional rack-and-pinion design) can be directly used to convert linear motion to rotational motion. For the sake of simplicity, it is assumed that a rack drives the pinion. With that established, it can be further stated that a rack moving linearly in the vertical plane and in mesh with a pinion rotating at a right angle to the linear motion can be used to convert the linear motion of the rack in the vertical plane to rotational motion of a shaft keyed to the pinion. Details of the geometry of the rack and pinion will be presented later in this chapter.

An underwater submerged buoyancy module or airlift bag or just an inflated bag attached to a rack (similar to that mentioned in the paragraph above) of negligible weight and in mesh with a pinion (underwater) can be shown to convert the buoyant forces acting on the bag, as it moves across the water column, to generate a torque on the shaft keyed to the pinion. This description is represented in Figure 12.5.

The buoyant forces acting on the bag can be utilised directly in calculating the power of the output shaft at the rack. Torque, τ, on the pinion due to the linear motion of the rack, is given by

$$\tau = b_g r \tag{12.2}$$

The effective buoyancy on the bag is assumed to be constant as it moves through the water column. The mathematical variation of buoyancy with water depth will be shown later. For a constant ascent speed of the bag, the rotational speed of the pinion depends on the radius r. The vertical distance covered by the rack is represented by S. When S_l is the distance covered by the rack in one revolution of the pinion with radius r_l, then the distance S_l is

$$S_l = 2\pi r_l \tag{12.3}$$

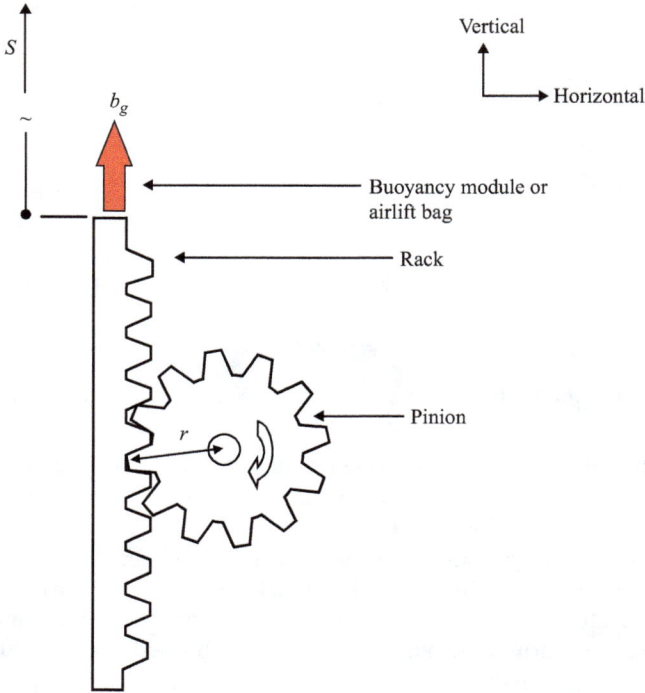

Figure 12.5   Buoyancy module or airlift bag or an inflated bag attached to a rack in mesh with a pinion

Assume another pinion with similar gear tooth geometry and half the radius of the pinion mentioned above (r_l), which is in mesh with the rack as it moves vertically. Hence the distance covered by the rack for every complete revolution of the smaller pinion is represented by

$$S_s = \pi r_s \tag{12.4}$$

Therefore, with every complete revolution of the larger pinion, the smaller pinion makes two complete revolutions.

The vertical distance covered by the rack per minute was given by

$$S_{\min} = 2\pi r \times RPM \tag{12.5}$$

The power P, due to the distance moved by the bag in 1 min, was calculated by

$$P = \text{buoyant force}, b_g \times \text{distance per minute}, RPM \tag{12.6}$$

This can be rewritten as

$$P = \tau \times 2\pi \times RPM \tag{12.7}$$

Equation (12.7) represents the power in the output-shaft keyed to the larger pinion in mesh with the rack. The power output from (12.7) is for an ideal case, where no forces act against the bag's buoyant force. For a seawater situation, the effective buoyancy of the inflated bag is reduced due to the resistant forces acting against it during its ascent.

12.2.4 Resistant forces

When the energy bag is rising across the water column under the influence of buoyancy, it is subject to resistance (drags), as shown in Figure 12.6, which reduces the net buoyancy of the energy bag. The resistive forces or drag due to the bag motion are the following:

1. Pressure drag or form drag.
2. Friction drag.
3. Gravity force.

12.2.4.1 Pressure drag

Pressure drag only applies to 'bluff bodies', bodies that are tall relative to their length [13]. Although the energy bag does not exactly fit into this category, it will be considered because the projected area of the bag is similar to a sphere with fairings (see Figure 12.7) and will be subject to a certain degree of pressure drag. The pressure acts perpendicular to every point on the bag's surface and varies along the surface of the bag – when undergoing vertical motion. Being greatest at the top, where the streamlines divide and least at the bottom where the streamlines are closest together (Bernoulli's equation). The build-up of positive pressure at the top and negative pressure at the bottom leads to a pressure imbalance which then causes a resultant pressure force.

The projected area of the bag represents the area perpendicular to the flow during ascent. The pressure force acting at any point of the small surface area of the bag was calculated by

$$F_P = P\, dA \tag{12.8}$$

Since the projected area of the bag is assumed to be spherical, the resultant pressure force was obtained by

$$F_P = \oint P \cos\theta\, dA \tag{12.9}$$

where dA represents small surface areas, the point where the pressure exerts the force. For simplicity, the pressure drag is calculated from the following equation:

$$F_P = C_P \frac{1}{2}\rho_w u^2 A \tag{12.10}$$

where C_P is the pressure drag coefficient.

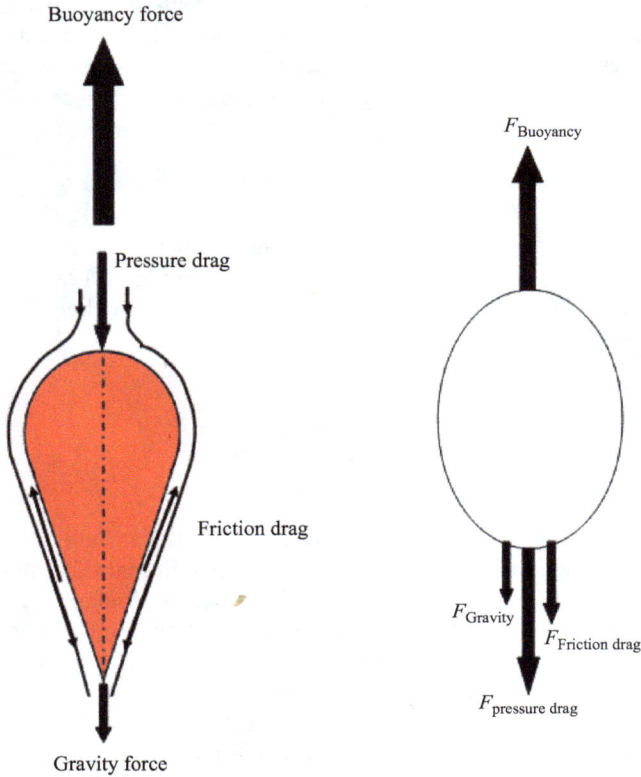

*Figure 12.6   Schematic showing the resultant forces acting on the energy bag at a
                constant velocity*

12.2.4.2   Friction drag

Friction drag is common with shapes with high length (L) to beam (B) ratios. Due to
the viscosity of seawater, a tangential force or skin friction is induced on the bag's
surface.

Figure 12.7 shows the components of the forces (drag) acting on the bag in the
direction of the velocity, u, while moving across the water column. Figure 12.8
shows a tiny patch (red colour – part of the inflated energy bag) moving vertically
through a seawater column at velocity u, and an imaginary stationary plate parallel to
the moving patch. As the bag moves through the water column, the shear force is
generated between the water molecules directly beside the patch and the moving
patch. This force, known as skin friction, results from the shearing resistance and
viscosity of seawater. The magnitude of the shearing force is derived by

$$\tau_{SF} = \frac{F}{A} \tag{12.11}$$

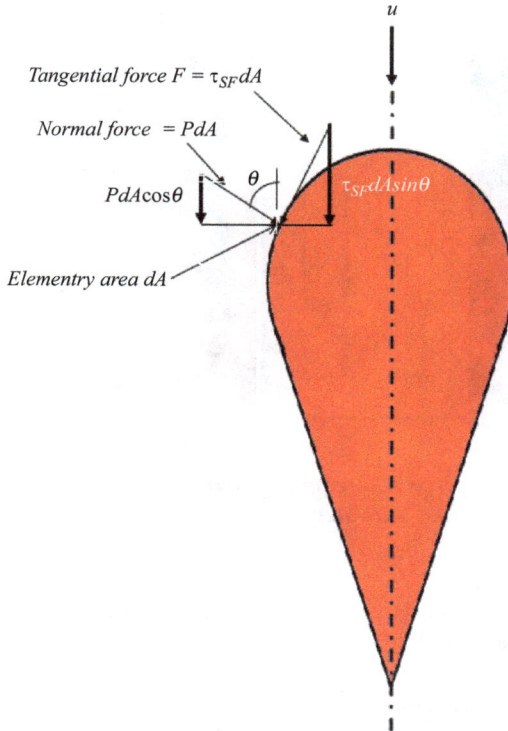

Figure 12.7   Components of drag on the energy bag

where A represents the area of the moving patch and F is the external force applied to the moving surface to overcome the resistance of the fluid acting tangential to the small patch, as shown in Figure 12.8. For simplicity, the friction drag was calculated as

$$F_F = C_F \frac{1}{2}\rho_w u^2 A \tag{12.12}$$

The total drag force acting on the bag is a sum of both the friction drag and the pressure drag and is calculated as

$$F_D = C_D \frac{1}{2}\rho_w u^2 A \tag{12.13}$$

where C_D represents the total drag force coefficient. It is derived experimentally, and it is a function of the shape of the body parallel to the fluid flow. Since the shape of a fully inflated and underwater submerged energy bag is similar to the

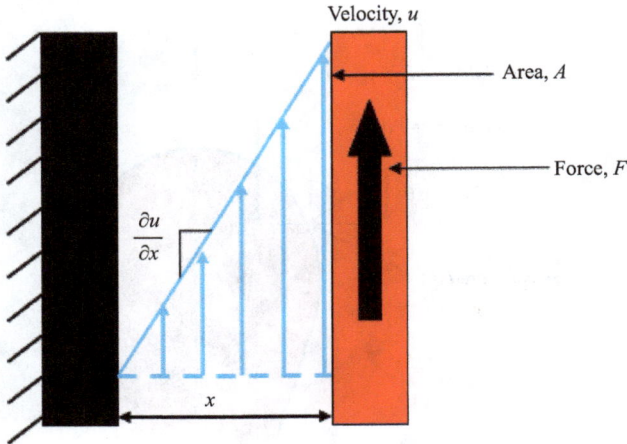

Figure 12.8   Viscous force acting against moving energy bag due to shear resistance of seawater

shape of a sphere with fairings, as shown in Figure 12.7, $C_D = 0.07$ [14,15] was applied to calculate total drag force.

12.2.5   Energy bag velocity and jerk (acceleration)

Jerk or jolt is a physics term representing the rate of change of acceleration with respect to time. The presence of the jerk value in the calculations made the solution cumbersome and complicated. To solve this, therefore, alternative considerations were made. Since the air density in the bag reduces with decreasing water depth, and the hydrostatic pressures reduce with decreasing water depth, there is a resultant volume increase on the bag as it rises to the surface. This increase in the bag's volume simultaneously increases the drag force and added mass acting on the bag. In this design, the steady increase of these forces is assumed to compensate for the steady proportional increase of the buoyant force. Based on this assumption, an approximation of a constant ascent velocity will be utilised.

The desired maximum ascent velocity of the energy bag is no more than 1 m/s based on the IMCA D 016 recommendations. As the pressure acting on the wall of the energy bag drops during ascent, the bag is expected to expand, still maintaining its geometric submerged shape of a sphere with fairings. The shape is maintained because the constant load from the gear system and generator are acting downwards (opposite the direction of ascent). Hence, the drag force coefficient, $C_D = 0.07$, is expected to remain constant throughout the ascent of the bag as the deformation of the bag was assumed to be negligible due to the low ascent velocity and subsequent low-pressure head [16] (this claim is based on theoretical investigations).

Figure 12.9 shows the forces acting on a sphere ascending under the influence of buoyancy at a velocity. This body will experience terminal velocity u_t when

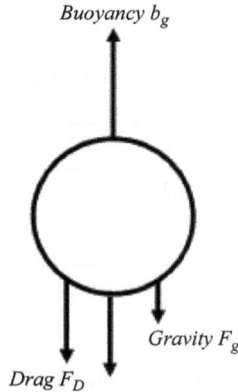

Figure 12.9 Free body diagram of the buoyancy generator (represented by a sphere)

$$b_g \approx F_g + F_D + F_{MR} \tag{12.14}$$

The mechanical resistance is as generated by the gear system (treated later in this chapter) while gravity, F_g, was calculated as

$$F_g = \left[\frac{4}{3}\pi r^3 \rho_a + \text{mass of bag}\right]g \tag{12.15}$$

The bag's volume can be obtained by applying the exact formula for calculating the volume of a sphere. It was assumed that the shape of the fully inflated energy bag before immersion in water is spherical.

The value of the mechanical resistance is dependent on the available force when the drag and gravity are subtracted from the effective buoyant force. That is, the gear system was designed based on the value of the available buoyant force. This value represents the minimum force required to generate useful power in the shaft. Therefore, the ascent terminal velocity u_t is gotten from

$$u_t = \sqrt{2\frac{\left(b_g - (F_g + F_{MR})\right)}{C_D \rho_w A}} \tag{12.16}$$

However, JW Automarine identified control of ascent velocity during lifting as a significant challenge [17]. This challenge is due to the expanding air during ascent and the resultant velocity head, which could lead to a high-pressure head, where the drag force acting on the bag becomes higher than the buoyant force – it is known to occur at very high velocities. It is again advised to maintain a velocity of no more than approximately 1 m/s [17]. Since the net force is not equal to zero, there will be an acceleration of the body as it moves through the water column; therefore, the added mass on the bag is not ignored. Where dampers are installed to

Table 12.1   The value of the change experienced on buoyant force,
drag force and mass of a charged energy bag across a
vertical water column of 100 m

	@ 500 m	@ 600 m	Change
Buoyant force (kN)	226.34	186.00	40.33
Drag force (N)	365.16	325.71	39.45
Mass (kg)	1,580.73	1,575.61	5.12

compensate for changes in the buoyant force, the added mass can be ignored. The added mass for a spherical body is given by

$$M_{\text{AVM}} = \frac{2}{3}\rho\pi r^3 \tag{12.17}$$

Table 12.1 shows the change in the buoyant force, drag force, mass and added mass of an energy bag as it rises 100 m through a water column [4].

Mechanical resistance of the gear system (gearbox) and other forces acting alongside it, such as the weight of the rack, will be treated in the next section. The drag due to the exposed gear teeth of the rack when moving across the water column will not be considered in this chapter and hence assumed to be negligible.

12.2.6   Transmission system

The derivation of (12.7) was based on the assumptions that a plane surface rack is moving relative to a plane (smooth) surface rotating pinion. That is, the resistance due to the rack-and-pinion gear teeth was not considered. Since the transferred torque and speed between gears depend on the number or arrangement of teeth of the gears, the gear ratio of the rack and pinion is considered.

12.2.7   Rack and pinion gears

A rack with n_r number of teeth per unit length in mesh with a pinion with n_p number of teeth can be assumed to be represented by the setup shown in Figure 12.10. If the angular displacement of the pinion is represented by θ, then one complete revolution of the pinion is given by

$$\theta = 2\pi \tag{12.18}$$

Similarly, a complete revolution of the pinion implies that n_p number of teeth of the pinion has made contact with the rack. Hence, the rack has moved a distance corresponding to n_p which is given by

$$S = \frac{n_p}{n_r} \tag{12.19}$$

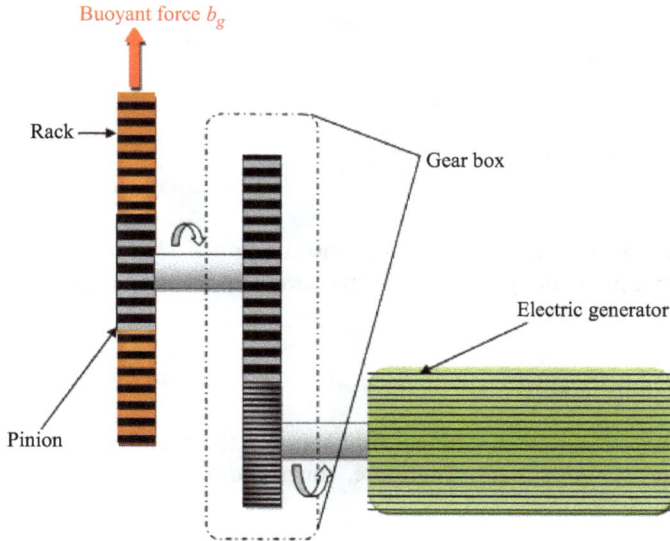

Figure 12.10 Schematics of DD, the power transmission system of the buoyancy engine

From (12.19), it has been shown that the displacement relationship between the rack and the pinion is a linear one and is seen to be exactly analogous to (12.3) and (12.4). That is, as the linear displacement of the rack increases, the angular displacement of the pinion increases proportionately; it is shown thus as

$$\theta = K.S \tag{12.20}$$

$$K = \frac{\theta}{S} = \frac{2\pi}{n_p/n_r} \tag{12.21}$$

Hence from (12.20), the displacement relationship between the rack and the pinion expressed in terms of the number of teeth of the gears is

$$\theta = 2\pi \frac{n_r}{n_p} S \tag{12.22}$$

The velocity relationship can be expressed in the same way as

$$\omega = 2\pi \frac{n_r}{n_p} u \tag{12.23}$$

where ω represents the angular velocity of the pinion while u represents the linear velocity of the rack. The translational power at a rack and the rotational power at a pinion are shown in (12.6) and (12.7). For a perfect system where there are no mechanical losses in the gear system, it is assumed that the translational power introduced into the rack-and-pinion system is exactly equal to the rotational power produced at the pinion shaft. That is,

$$\text{Translational power in} = \text{rotational power out} \tag{12.24}$$

Therefore, it is represented by

$$b_g u = \omega \tau \tag{12.25}$$

Based on this perfect system and substituting (12.23) into (12.25), the relationship between the buoyant force and the torque in the pinion expressed in terms of the number of teeth is given by

$$b_g = 2\pi \frac{n_r}{n_p} \tau \tag{12.26}$$

As shown in [4], the average efficiency of the traditional rack-and-pinion system is about 94 per cent. Therefore, for a traditional design rack-and-pinion system, (12.25) becomes

$$b_g u 0.94 = \omega \tau \tag{12.27}$$

12.2.7.1   Gear mesh geometry

The optimum mesh geometry of the rack and pinion gears of the buoyancy engine is a rack with an infinite number of teeth and a pinion with a radius large enough to generate maximum torque. The mesh between the gears primarily looks to transfer force and linear velocity (IN) from the rack to achieve maximum torque and angular velocity (OUT) at the pinion and ultimately power at the shaft. In selecting the number of teeth and radius of the pinion, it was noted that hydrodynamic drag would be generated between its teeth during rotation. The magnitude of the drag increases with the pinion's size and the number of teeth. This parameter will be ignored and not considered in this chapter as roller pinion gears will be utilised. The speed-increasing gearbox will be fully maximised in stepping down the input torque to the generator rated torque and speed.

In selecting the mesh geometry for this rack-and-pinion gear system, excessive clearance is advantageously avoided. Conventional rack and pinion gears work in environments where the ambient temperature could be raised to values higher than the allowable working temperature of the gear and hence cause expansion of the gear system and ultimately lead to a jam of the whole system. However, the buoyancy engine's rack-and-pinion gear system works in environments that have an almost constant and low ambient temperature of about 5 °C; therefore, expansion due to the temperature rise of the gear is eliminated. Therefore, the clearance in the mesh geometry is removed; hence backlash and overall system efficiency losses are ultimately avoided. A rack-and-pinion system without backlash can convert rotary motion to linear motion with an efficiency of 99 per cent and an accuracy of ±20 μ regardless of the travel length [18]. This chapter will assume that this efficiency is exactly reversible; hence, 99 per cent is taken for conversion of linear motion to rotary motion by the rack and pinion.

12.2.8   Gearbox

Gearboxes can generally be described as a machine that transmits power from one point to the other with the primary purpose of varying the input torque and speed from the output torque and speed. For the buoyancy engine, the sole purpose of a gearbox is to convert the slow, high-torque angular velocity of the pinion shaft to a much higher angular velocity and a lower toque to match the operational require-ments of the electric generator. This standard requirement of the gearbox is exactly analogous to the function of gearboxes in wind turbines. Studies on the efficiency and effectiveness of gearbox-fitted wind turbines showed that gearboxes are the singular most expensive component of the wind turbine (10 per cent of turbine + installation cost). It is characterised by high failure rates far earlier than the design life of the turbine [19,20]. Other problems associated with these gearboxes include their need for constant lubrication and hence incorporate complex electronics for monitoring and diagnosing failure that could lead to the failure of other sensitive components.

Siemens, one of the biggest manufacturers of wind turbines, especially gearbox-fitted turbines, has carried out extensive research and testing on direct-drive (DD) wind turbines [21]. DD wind turbines are designed with no gearbox, and the rotor is directly connected to a low-speed multi-pole generator which generates at the same speed as the rotor. Although this technology has been used and developed by other wind turbine manufacturers such as Germany based Enercon and Vensys as well as ScanWind – Norwegian origin, now acquired by General Electric [19].

These findings instructed the abrupt change in the initial plan to utilise a buoyancy engine drive-train gearbox. Furthermore, because all subsea installations must be designed to function throughout their design life as any form of major maintenance is a costly option, the buoyancy engine was designed without a gearbox. *'It is a secure hard fact that if you do not have a component, you do not have any maintenance or failures associated with it'*, by Jamieson [20].

This change implies that the pinion gear geometry, torque and velocity are selected based on the operating parameter of the permanent magnet generator (PMG) – low-speed multi-pole generator. Figure 12.11 shows a schematic diagram of the DD buoyancy engine. This diagram represents only a section of the actual design as the multiple pinions in mesh with the rack has been deliberately omitted.

It is important to note that the major limitation of the DD system, as experi-enced by the wind turbine industry, is the challenge of very high loads on the shaft bearings. Other common problems with this system include the need for complex designs and inaccessibility of some vital components [19].

12.2.9   Power generator

The choice of power generator to be utilised in the system is the PMG due to the findings stated in the previous section.

PMGs have 100 or more magnetic pole pairs for a range of speed of about 6–18.5 revolutions per minute (rpm). In contrast, conventional generators have

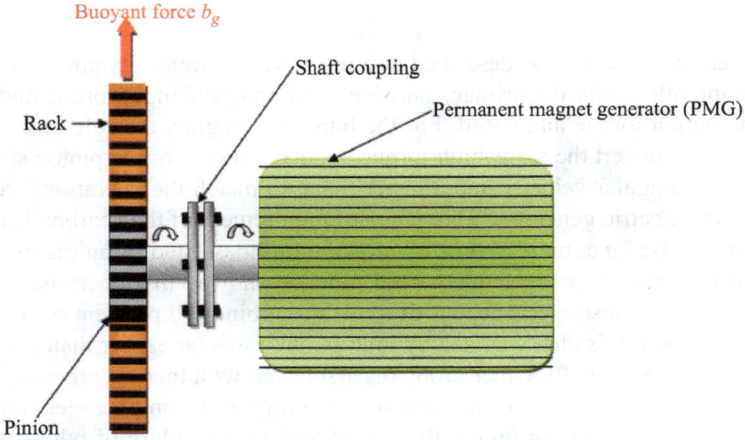

Figure 12.11   Schematics of DD/gearless system with a PMG

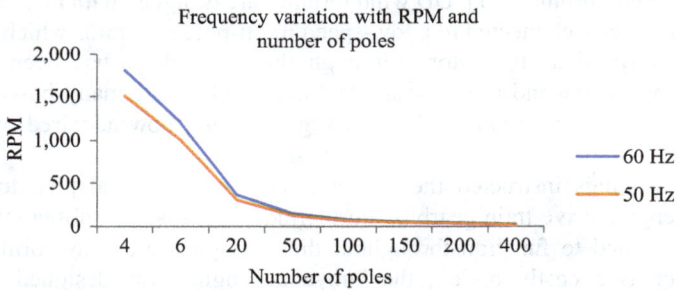

Figure 12.12   Relationship between the frequency, rpm and the number of poles of the generator

about 4–6 magnetic pole pairs (50 and 60 Hz, respectively) for a speed of about 1,800 rpm (4 pole – 60 Hz) [19,22]. The relationship between the number of poles *n*, the generator speed in *rpm* and frequency *f* is

$$f = n \frac{rpm}{120} \quad [\text{Hz}] \tag{12.28}$$

The generators speed in rpm is expressed as

$$\text{Generator speed [rpm]} = 120 \frac{f[\text{Hz}]}{n} \tag{12.29}$$

Increasing the number of poles on the generator allows it to rotate at lower speeds for constant frequency output. This trend is as plotted in Figure 12.12.

Table 12.2   Trend of speed for varying number of poles of a generator

Magnetic pole pair	60 Hz	50 Hz
4	1,800 rpm	1,500 rpm
6	1,200 rpm	1,000 rpm
100	72 rpm	60 rpm

Table 12.3   The trend of the value of gearbox ratio of a system for a constant frequency and pinion speed

Number of poles	4	6	20	50	100	200	400
Generator (rpm)	1,800	1,200	360	144	72	48	18
Pinion (rpm)	18	18	18	18	18	18	18
Gearbox ratio	100.00	66.67	20.00	8.00	2.67	2.00	1.00

When there is no direct coupling between the pinion shaft and the generator rotor and they are linked by a gear system, the gear ratio, G, between them is defined by

$$G = \frac{\text{Speed of generator [rpm]}}{\text{Speed of opinion [rpm]}} \tag{12.30}$$

From (12.29) and (12.30), it is shown [4] that for a given buoyancy engine system at a constant rack linear velocity (or constant pinion rpm), increasing the number of poles of the generator (Table 12.2) in (12.29) reduces the gearbox ratio required between the pinion and the rotor – for a system fitted with a gearbox. Table 12.3 shows this trend for a buoyancy engine with a constant pinion speed of 18 rpm and an expected output frequency of 60 Hz. It can be seen that the need for a gearbox diminishes as the number of magnetic pole pairs of an electric generator increases. The parameters for which the gearbox ratio is unity are for the conditions when the pinion speed matches the generator speed at a constant frequency of 60 Hz. Based on the trend shown in Table 12.3, the selection of the optimum generator is primarily dependent on the power in the output shaft of the pinion gear. A gearbox is not needed where the ratio is unity.

12.3   A model plant

The method used in determining the optimum parameter for this design is based on mathematical models and the formulas previously presented in this chapter. This chapter did not consider the design of the energy bag and the optimisation of the quantity of storage air for given power output. Hence the optimum volume of air stored within a bag for given power output and at the given water depth have been adopted from the calculations carried out by Seamus [2]. Considering the IMCA D 016 regulations, the energy bag is filled 100 m below the depth at which the air will be released with the equivalent amount of air required at the release depth. That is, the known or optimum volume of air required for a specific water depth as gotten from [2] will be scaled to fit into parameters at a depth 100 m greater with no additional air added to or removed from the system. Hence, the bag rises a distance of 100 m to the optimum without adding destructive stresses on the bag membrane.

12.3.1   Buoyant force

In [2], it is shown that at 1 GW (86.4 TJ/day), the required volume and mass of air is 1,488,300 m^3 and 130,550 tonnes, respectively, at seawater ambient temperature and pressure ratio of 70 with the atmosphere. These selected optimum parameters are utilised to design the buoyancy module and determine the forces acting on the bag as it moves across the water column.

Table 12.4 shows the results of inputting the above parameters into relevant formulas. The values obtained at the water depth of 500 m were subjected to the higher hydrostatic forces acting at 600 m in order to find the scaled-down parameters. For better understanding, assume that when the bag is filled to its optimum volume at a water depth of 500 m, it is sealed and dragged down to a depth of 600 m without adding or taking out air from it. It can be imagined that due to the increased hydrostatic forces, the volume of the bag will be decreased, and the density of the air will be increased. To mathematically show this, Boyle's law was applied for the closed system [4]. The results shown in Table 12.4 imply that when the bag is positioned at a water depth of 100 m higher than the optimised, the bag needs to be filled to a volume of 1,305,674.057 m^3. Hence, when the bag eventually rises by 100 m to the optimised depth, the bag will return to the optimised volume of 1,488,300 m^3.

Based on this finding, the need to release stored air according to the IMCA D 016 regulations will not be necessary. However, it is necessary to fix a pressure release valve to the bag for emergency – like a burst or escape.

Based on these results, the energy bag, when filled with air, will generate a gross buoyant force of 11,821 MN, which is an estimated value of the total upward force. To determine the net translational force exerted by the bag on the rack, the drag forces acting on the bag will be subtracted from the total buoyant force.

Table 12.4   Optimum volume of air for pressure compensation during ascent of bag

	Bag volume (m^3)	Air density (kg/m^3)	Buoyant force (MN)
@ 500 m	1,488,300.000	90.166	13,677
@ 600 m	1,305,674.057	102.778	11,821

12.3.2   Resistant forces

When the energy bag rises along the water column under the influence of buoyancy, it is subject to resistance (drag), which reduces the effective buoyancy of the energy bag. The major resistant forces acting on the bag, as earlier mentioned, are drag forces, gravity force, mechanical resistance and resistance due to the added mass when there is an acceleration of the bag.

The desired maximum ascent velocity of the energy bag is 1 m/s; hence, the bag is expected to grow bigger but maintain its geometric submerged shape as the deformation of the bag is assumed to be negligible due to the low ascent velocity and subsequent low-pressure head.

12.3.2.1   Drag force

The established formula (12.13) for deriving the total drag forces (friction and pressure drag) acting on the body is utilised to calculate the total drag forces acting on the bag during its motion. The recommendation given in the JW Automarine lift bag operational guide [17] was employed in selecting of the ascent velocity of the bag to minimise the drag forces on the bag and avoid the problems of damage to the bag due to the pressure head. Based on this, the drag force on the bag for the various values of ascent velocities (no more than 1 m/s) was considered in calculating the optimum velocity. Ascent duration and the power to the system were considered in the calculation of the drag forces.

To determine the surface area of the energy bag, the bag was treated as a sphere since the shape of the energy bag before immersion in water is similar to that of a sphere. The known volume of the bag was used to determine its radius. The derived radius was then used to calculate the area of the drag head. The area was calculated to be 14,441.4 m^2. Drag forces for various velocities from 0.1 to 1 at intervals of 0.1 was calculated [4], and the results are shown in Table 12.5 and Figure 12.13.

Table 12.5   Various ascent velocities of the bag across the water column and its corresponding drag force

Terminal velocity (m/s)	Drag force (kN)
0.1	51.81
0.2	103.62
0.3	155.43
0.4	207.23
0.5	259.04
0.6	310.85
0.7	362.66
0.8	414.47
0.9	466.27
1.0	518.08

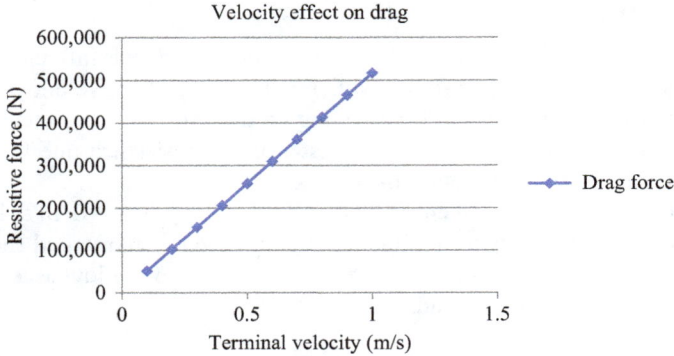

Figure 12.13   Effect of the ascent terminal velocity on the total hydrodynamic forces acting on the bag

Table 12.6   Gravity forces acting on the energy bag at two chosen distances along the axis of ascent

Travel/ascent distance of bag (m)	Mass of bag (kg)	Mass of air (kt)	Gravity force (MN)
0	2,658.83	134.18	1,317.24
50	2,658.83	125.94	1,236.42

12.3.2.2   Gravity force

The gravitational force acting on the bag was calculated from (12.15). The weight of the bag in water was ignored in this chapter due to insufficient information on the components of the bag. The mass of the empty energy bag was used as a constant to calculate the gravitational force acting on the filled bag. The mass of the air was calculated for two submerged positions of the bag: first, when the ascent distance is zero, and then when the ascent distance is at the midpoint, that is, 50 m.

The mass of the bag was obtained by iteration, using the known masses of two existing prototype energy bags (used for test study of the ICARES in the University of Nottingham) of different sizes and diameter [8]. The dimensions after iteration are given as follows:

1.   Diameter, 1.8 m; total mass, 12.10 kg
2.   Diameter, 5 m; total mass, 75.4 kg
3.   Diameter, 135.6 m; total mass, 2,658.83 kg – *after iteration*

For the two points for which the gravitational forces were calculated, it can be seen that the values of the gravitational forces acting against buoyant forces tend to reduce with increasing water depth (Table 12.6).

The resistant forces acting against the bag's buoyancy were all calculated except for the added-mass effect and the mechanical resistance generated in the generator. The added mass on the bag was deliberately omitted because the acceleration only lasted few seconds at the beginning of the ascent. This assumption is so because this machine attains full load in few seconds, which induces *full-load torque* (torque needed to continue to generate the rated power at high loads on a generator) *resistance* in the generator. The calculation of the torque resistance induced by the generator is not shown in this chapter, but it is considered to be sufficient enough to compensate for any effect of an increase in the buoyant force acting on the bag as it ascends to lower pressures. When the rated design speed is attained at full load on the generator, the full-load torque is induced, along with its resistance. The mechanical input power, P_{IN} required (from the prime mover) to overcome the full-load torque [13] is given in (12.31) by the sum of the following:

1. Core loss power $P_{core\ loss}$
2. Electrical loss power $P_{elec\ loss}$ – often negligible
3. Output power P_{out}
4. Mechanical loss power $P_{mech\ loss}$

$$P_{IN} = P_{core\ loss} + P_{elec\ loss} + P_{out} + P_{mech\ loss} \qquad (12.31)$$

Since the effect of a change in the buoyant force of the bag is cancelled by the corresponding increase of the full-load torque resistance (rotational loses) from the generator, the system will only experience acceleration for few seconds at the beginning of the ascent. After the short period of acceleration has elapsed, the bag will continue to rise at a terminal velocity. Based on theoretical investigations, the total resistive forces on the system determined the value of the terminal velocity.

12.3.2.3 Added mass

Since it has been established that the energy bag will be subject to negligible acceleration and will ascend at a terminal velocity which is dependent on the resistive load, therefore, the added mass on the bag during its ascent was ignored.

12.3.3 Net buoyant force

The resultant or net buoyant force, $b_{g\ net}$, acting on the bag is derived from

$$b_{g\ net} = b_g - (\text{gravity force } F_G + \text{drag force } F_D) \qquad (12.32)$$

Based on the gross buoyant force calculations, the value obtained at a water depth of 600 m was utilised to determine the net buoyant force on the bag. This is so because the buoyant force at a depth of 600 m is the closest reference when determining the tangential force required for overcoming the sum of the friction torque, the breakaway torque and the acceleration torque.

Table 12.7   Net buoyant forces acting on the energy bag during ascent at various values of ascent velocity

Velocity (m/s)	Drag force (kN)	Gravity force (MN)	Buoyant force (MN)	Net buoyant force (MN)
0.1	51.81	1,236.42	13,677	12,440.53
0.2	103.62	1,236.42	13,677	12,440.47
0.3	155.43	1,236.42	13,677	12,440.42
0.4	207.23	1,236.42	13,677	12,440.37
0.5	259.04	1,236.42	13,677	12,440.32
0.6	310.85	1,236.42	13,677	12,440.27
0.7	362.66	1,236.42	13,677	12,440.21
0.8	414.47	1,236.42	13,677	12,440.16
0.9	466.28	1,236.42	13,677	12,440.11
1.0	518.09	1,236.42	13,677	12,440.06

The value of the gravity force at an ascent height of 50 m was utilised in the calculation of the net buoyant force because it acts as an average. Also, in calculating the mass of the bag components, the weight in water was not considered. The drag forces for velocities from 0.1 to 1 m/s (0.1 m/s increments) are calculated and the optimum value is selected [4].

The net buoyant force due to the difference in the density of the overall energy bag and the volume of water displaced during ascent is shown in Table 12.7. The calculations are based on the established parameters.

The optimum value of the net buoyant force is equal to the translational force on the rack, which is also the tangential force acting on the pinion.

12.3.4   Transmission system

The optimum net buoyant force is equal to the translational force on the rack, and it is the same as the tangential force on the pinion. The selection of the optimum value of the translational force and velocity of the rack is mainly influenced by the speed and power requirement of the electric generator. This is so because power is the product of force and velocity; therefore, lower velocity values will result in lower values of power. However, this chapter earlier established that the rotational speed of DD multiple pole pair generators (in the megawatt range) could operate in the range of 6–18 rpm [22]. Therefore, the selection of the optimum rack velocity is greatly influenced by the generator speed. The upper limit value of the gear speed is used in selecting the optimum gear ratio of the rack and pinion gears. In order to transfer the maximum transferable value of the power at the rack to the pinion gear, and maintain the rated speed of the generator, a pinion speed of 0.3 revolutions per second (rps) was selected, which means that it makes one complete revolution in 3.333 s. An assumed initial value of the rack's radius was selected to solve this problem. A radius of 1 m was the initial selection. This value was later varied at intervals of 0.1 m down to 0.4 m [4]. The calculation was made based on (12.3), and the results are as presented in Table 12.8 for a constant pinion speed of 0.3 rps.

Table 12.8   The influence of the pinion diameter on the rack translational velocity for a constant velocity of the pinion

Pinion radius (m)	Pinion speed (rps)	Rack speed (m/s)
1.0	0.3	1.89
0.9	0.3	1.70
0.8	0.3	1.51
0.7	0.3	1.32
0.6	0.3	1.13
0.5	0.3	0.94
0.4	0.3	0.75

The result (for values of the pinion radius greater than 0.6 m) can be seen to go against the recommendation of the JW Automarine [17] that the ascent speed of a lift bag must never exceed 3 feet per second (~1 m/s). Figure 12.14 shows four pinions and when all four pinions are imagined to rotate at the same angular velocity (rpm), the rack in **A** will move at a higher linear velocity than the rack in **B**, **C** and **D** in that sequence. The optimum value of the pinion radius must be a value less than 0.6 m.

The optimum velocity of the rack is utilised in calculating the power at the rack, due to the buoyant force, and the resultant torque and power at the pinion based on the efficiency of the rack and pinion gears. Equation (12.25) is used to achieve this.

Based on the power attained in the pinion, the electrical power output of the generator is determined after the safety factor has been considered.

12.4   Optimisation and conclusion

Most of the parameters to be optimised in this chapter have already been presented in Model Plant above. Hence, this section collates the optimised results attained from the design of a buoyancy engine model plant presented in this chapter.

12.4.1   IMCA D 061 regulation

The IMCA D 061 regulation for using closed bags for lifting involving ascent was considered and a mitigation technique was employed to avoid associated problems. The mitigation technique employed includes the following:

1. For air stored at a depth 100 m greater than the air extraction depth, Boyle's law was used to determine the optimum volume of air required at the extraction depth. The optimised volume was then pumped into the bag at a depth 100 m greater than the optimised.
2. Design the bag with a higher factor of safety (greater than 5).
3. For unavoidable emergencies, a pressure relief valve is required on the energy bag.

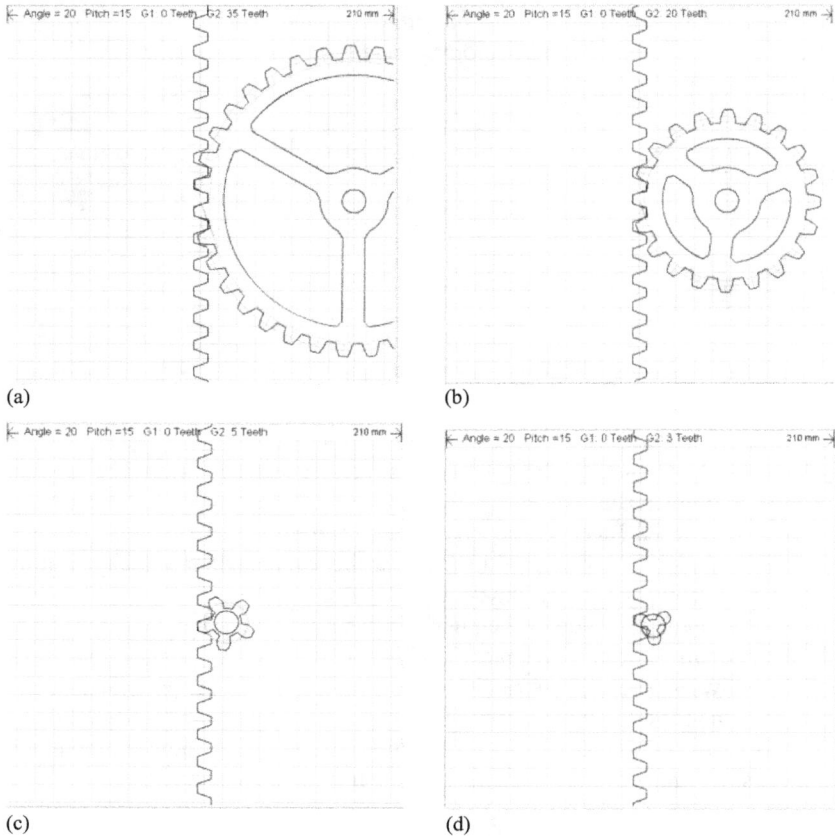

(a)

(b)

(c)

(d)

Figure 12.14   Rack in mesh with pinions of different radius rotating at similar angular velocity

For this design, the volume of the energy bag (multiple) for 1 GW (turbine power) at the extraction depth is 1,488,300 m². The buoyancy at this stationary position is 13,677 MN, which is 100 m below the air extraction depth. The bag's volume is 1,305,674.057 m³, and the buoyant force at this stationary position is 11,821 MN.

12.4.2   Added mass

The added mass in the system has been ignored because resistance due to full-load torque is assumed to compensate for the extra buoyant force generated due to an increase in volume as it rises to lower pressures. These have been extensively discussed in the model plant section.

12.4.3   Gravity force

The gravity force acting on the bag was calculated at two points along the ascent path, when the ascent distance is zero and when ascent distance is 50 m. To determine the bag's weight, an iteration of two different bags of different sizes was made. The value of the gravity force when the ascent height is 50 m was utilised because the weight in water of the component materials of the bag was ignored. The value of the gravity force utilised is 1,236.42 MN.

12.4.4   Drag force

Velocity was found to have the highest influence on the total drag force acting on the body. The recommendation given in the JW Automarine lift bag operational guide was employed; hence, the ascent velocity of no more than 1 m/s was used to calculate the drag force, as seen in Table 12.5. The drag force involved making some assumption, one of which is that the bag's shape before being submerged is spherical. Therefore, using the known volume, the radius is determined. Varying values of drag force were determined for different velocity values.

12.4.5   Net buoyant force

The net buoyant forces acting on the energy bag were calculated from (12.33) and are shown in Table 12.7. The net buoyant force values were derived because they directly depend on the drag force and indirectly depend on the ascent velocity. The optimum net buoyant force is dependent on the optimum rack dimension of the corresponding speed. This means that if the optimum net buoyant force is at a value corresponding to 0.6 m/s on Table 12.7, then it must correspond to a rack design of 0.6 m/s.

12.4.6   Transmission system

The selection of the translational force and velocity of the rack is mainly influenced by the speed and power requirement of the electric generator. Based on existing DD multi-polar pair generators, the speed often falls within 6–18 rpm. The upper limit of the speed was utilised to derive the corresponding pinion speed per second –0.3 rps. An assumed radius was derived to back solve with which a range of rack speed emerged. Most of the rack speed values exceeded the recommended speed of ascent by JW Automarine [17]. These figures are presented in Table 12.8, and the values of the pinion radius greater than 0.6 m generate rack speed above what is recommended.

12.4.7   Optimum power

From the results shown in Table 12.8, only two values of the rack speed (0.942 and 0.754 m/s) fall below the maximum speed threshold of 1 m/s. These parameters were then used to find the value of net buoyant force, which ascends at a velocity equal or nearest to the velocity of the rack. This was done for the two values of the rack speed that fall within the allowable. The results are shown in Table 12.9. For a given rack speed, a corresponding pinion with an optimum radius will rotate at 0.3 rps.

Based on these parameters, (12.25) was applied in determining the torque at the pinion as well as the power input to the system. Since the efficiency of the chosen rack and pinion gear to be used (Nexgen roller pinion system) [4,23] is known, at 99 per cent, then the efficiency was inputted into the equation to determine the exact value of the power at the pinion. Therefore, (12.33) for the efficiency of 99 per cent is rewritten as

$$b_{g_{NET}} u 0.99 = \omega \tau \tag{12.33}$$

The torque at the pinion for the corresponding buoyancies and rack speed are shown in Table 12.9.

These values of the pinion torque are directly used to determine the power on the shaft connected to the generator [4]. The two values of shaft power determined for the two rack speeds are enumerated in the following. The lower value of the power was selected to arrive at an optimum value.

- Power in the pinion $P \approx 9.29 \text{GW}$

Since the selected ascent speed is 0.754 m/s, the total ascent time through the ascent distance of 100 m was calculated to be 133 s (2.2 min). Hence, the total output power for the period of ascent is approximately 20.43 TJ.

12.4.8   Conclusion

The biggest challenge of the CAES concept lies in making it more efficient, increasing the energy output capacity, dissociating it from fossil fuels and making it easier to install. Compared to existing diabatic CAES plants, an UW-ACAES with a buoyancy engine system can deliver similar energy outputs at better environmental efficiencies. Based on the numerical results obtained from the optimisation of the buoyancy engine, the electrical power generated by the electric generator in the bulb turbine can be effectively applied in producing multiple electrical arcs in a purpose-built heat exchanger system. The short-term electrical energy is sufficient in maintaining the electric arc to generate sufficient sensible heat and stored in charged molten salt at over 500 °C. The molten salt can then be utilised to expand the stored air through multiple turbines or employed to produce superheated steam for turbine generators similar to those used in any conventional coal, oil or nuclear power plant [2].

Based on the parameters utilised, optimum power obtained from the system shows that the buoyancy engine is a very suitable means to generate the clean external heat of expansion required by UW-ACAES system to boost their power output and overall plant efficiency.

Table 12.9   Optimum operating parameters for the rack-and-pinion gear system

Rack speed (m/s)	Energy bag speed (m/s)	Pinion speed (m/s)	Net buoyancy (MN)	Pinion torque (MN-m)
0.94	0.9	0.3	12,440.11	3,867.13
0.75	0.8	0.3	12,440.16	3,095.36

References

[1]   University of Nottingham. Energy bags and supper batteries – Researchers develop innovative energy storage solutions. The University of Nottingham, 2008. [Online]. www.nottingham.ac.uk/news/pressrelease/2008/june/energybagsandsuperbatteries.aspx.

[2]   Garvey, S.D. *The dynamics of integrated compressed air renewable energy systems*. Nottingham: Science Direct, 2012. Vol. 39.

[3]   Thompson, G. [The King of Random]. (2015, 12 Feb). *Mini Arc Furnace (Arc Reactor Technology IRL)* [Video]. YouTube. https://www.youtube.com/watch?v=JIlZsuRc9jQ

[4]   Onwuchekwa, D.C. *Design of offshore renewable energy device: Compressed air energy storage*. Glasgow: University of Strathclyde, 2013.

[5]   Abshire, L.W., Desai, P., Mueller, D., Paulsen W.B., Robertson, R.D.B., Solheim, T. Offshore permanent well abandonment. Silo.Tips. Schlumberger, 2012. [Online]. https://silo.tips/download/offshore-permanent-well-abandonment.

[6]   The International Marine Contractors Association. Underwater air lift bags. www.IMCA-int.com. [Online]. http://www.imca-int.com/media/70860/imcad016.pdf.

[7]   Mana, D.S.K., Gourvenec, S., Hossain, M.S., Randolf, M.F. *Experimental investigation of the undrained response of a shallow skirted foundation subjected to vertical compression and uplift*. Rotterdam: American Society of Mechanical Engineers, 2011. OMAE2011-49072.

[8]   Pimm, A.J. *Analysis of flexible fabric structures*. Nottingham: University of Nottingham, 2011.

[9]   Merritt, H.E. *Gears*. London: Sir Isaac Pitman & Sons Ltd., 1962.

[10] Ewert, R.H. *Gears and Gear Manufacture – The Fundamentals*. New York: International Thomson Publishing, 1997. 0-412-10611-6.

[11] Rao, T.G., Saha, S.K, and Kar, I.N. *Sensor-actuator based Smart Yoke for a rack and pinion steering system*. India: *Mobility Conference on Emerging Automotive Technologies*, 2008.

[12] Shute-Upton Engineering. Rack and pinion. Shute-Upton Engineering. [Online]. http://www.shute-eng.com.au/gear_types/racks_pinions.html.

[13] Boundary layer theory. Freestudy.co.uk. [Online]. www.freestudy.co.uk/fluid%20mechanics/t3203.pdf.

[14] Benson, T. Shape effects on drag. http://www.nasa.gov. [Online] 62014. http://www.grc.nasa.gov/WWW/Wright/airplane/shaped.html.

[15] IIT Kanpur. *Control of Boundary Layer Separation*. [book auth.] Indian Institute of Technology. Mechanical Engineering Fluid Mechanics. s.l.: IIT Kanpur NPTEL Online, 2020.

[16] Onwuchekwa, D.C. *Offshore renewable energy storage: CAES with a buoyancy engine*. France: IEEE Oceanic Engineering Society, 2019.

[17] JW Automarine. Underwater lifting bag manual. JW Automarine, 2020. [Online].http://jwautomarine.co.uk/download/lb_man.pdf.

[18]   Nexen Europe. Linear motion control. Nexen Europe Group, 2013. [Online]. [Cited: 8 20, 2013.] http://www.nexeneurope.com/en/linear-motion-control.php.

[19]   Ragheb, A.M., Ragheb, M. Wind turbine gearbox technology. In: *Fundamentals and Advanced Topics in Wind Power*. INTECH. [Online] INTECH, 2011. [Cited: 8 21, 2013.] http://cdn.intechopen.com/pdfs/16248/InTech-Wind_turbine_gearbox_technologies.pdf. ISBN: 978-953-307-508-2.

[20]   Aubrey, C. Wind turbines – Direct drive options challenge gearboxes. *Wind Energy and Electric Vehicle Review (REVE)*. [Online] 10 31, 2010. [Cited: 8 21, 2013.] http://www.evwind.es/2010/10/31/wind-turbines-direct-drive-options-challenge-gearboxes/8100.

[21]   Stiesdal, H. Wind Power. Blazing the trail for offshore wind. s.l.: Siemens, 2013.

[22]   ENERCON GMBH. Product details on ENERCON wind turbines ranging from 800–7,580 kW. ENERCON. [Online]. http://www.enercon.de/en-en/64.htm.

[23]   Kliber, T. Design world: Rolling on down the line. Nexen. [Online] Nexen Group, 11 16, 2012. [Cited: 8 20, 3013.] http://www.nexengroup.com/nxn/company/news/id/85.

Index

www.ingramcontent.com/pod-product-compliance
Lightning Source LLC
Chambersburg PA
CBHW050512190326
41458CB00005B/1514